AF598018

Methods in Molecular Biology

Series Editor
John M. Walker
School of Life Sciences
University of Hertfordshire
Hatfield, Hertfordshire, AL10 9AB, UK

For further volumes:
http://www.springer.com/series/7651

The TNF Superfamily

Methods and Protocols

Edited by

Jagadeesh Bayry

INSERM Unité 1138, Paris, France

Editor
Jagadeesh Bayry
INSERM Unité 1138
Paris, France

ISSN 1064-3745 ISSN 1940-6029 (electronic)
ISBN 978-1-4939-0668-0 ISBN 978-1-4939-0669-7 (eBook)
DOI 10.1007/978-1-4939-0669-7
Springer New York Heidelberg Dordrecht London

Library of Congress Control Number: 2014937463

Printed on acid-free paper

Humana Press is a brand of Springer
Springer is part of Springer Science+Business Media (www.springer.com)

Preface

Since the documentation of TNFα and β 30 years ago, the TNF superfamily has witnessed a steady expansion of its members. Currently in humans, 19 ligands and 29 receptors of the TNF superfamily have been identified, and 3 additional receptors are present in mice. The major TNF superfamily ligands include TNFα, TNF β, TWEAK (TNF-related weak inducer of apoptosis), TL1A (TNF-like ligand 1A), OX40L, LIGHT (CD258, homologous to lymphotoxins, exhibits inducible expression, and competes with HSV glycoprotein D for HVEM, a receptor expressed by T lymphocytes), GITR (glucocorticoid-induced TNF receptor family related gene) ligand, Fas ligand, CD40 ligand (CD154), CD30 ligand, CD27 ligand (CD70), BAFF (B cell activating factor) or BlyS (B lymphocyte stimulator), and APRIL (a proliferation-inducing ligand). The main TNF superfamily receptors include Fas (CD95), CD40, CD27, CD30, OX40, DcR3 (Decoy Receptor 3), DR3 (death receptor 3), GITR, BAFF receptor, BCMA (B cell maturation), RANK (receptor activator of NF-kB), TACI (transmembrane activator and CAML interactor), TNF Receptor (TNFR) I and II, and CD137.

The majority of TNF superfamily members are type II transmembrane proteins. TNF family members have diverse biological functions: providing signals for activation, differentiation, survival and death of cells, modulation of immune response and inflammation, hematopoiesis, and osteoclastogenesis. Therefore, TNF family member proteins are implicated in the pathogenesis of several autoimmune and inflammatory diseases, cancer, and graft-versus-host disease and hence are viable targets to treat these diseases. In fact, TNF-blocking therapies such as monoclonal antibodies to TNFα (Infliximab, Adalimumab) or dimeric fusion protein consisting of the extracellular ligand-binding portion of the human 75 kDa (p75) tumor necrosis factor receptor (TNFR) linked to the Fc portion of human IgG1 (Etanercept); monoclonal antibodies to BLys (Belimumab) are already in clinics to treat diverse inflammatory conditions.

This volume of *Methods in Molecular Biology* focuses on various aspects of the TNF Superfamily members in health and disease. The Chapters 1–6 mainly present protocols to understand the signaling process of TNF family members. Chapters 7–10 provide technical examples of investigating the role of TNF family members in physiopathologies. Chapters 11 and 12 delineate the protocols on modulation of TNF signaling by pathogens. Chapter 13 outlines the experimental applications of TNF-reporter mice. The rest, Chapters 14–18, offers methodologies for various assays of TNF family members and the production of recombinant molecules.

I should honestly admit that this volume might not provide complete coverage of all TNF family members. As the TNF superfamily is too diverse and has such a wide range of implications in physiopathologies, it would require many volumes to cover all of them satisfactorily. In this volume, I have tried to provide a glimpse of the versatility of these members. I hope that the scientific community working in this field will benefit from the protocols presented in this volume. I am grateful to all the contributors for sharing their time and laboratory experience in the form of protocols, my family, and to John Walker, the series editor, for his advice, support, and inspiration.

Paris, France *Jagadeesh Bayry*

Contents

Contributors

RAFIA S. AL-LAMKI • *Department of Medicine, University of Cambridge, Cambridge, UK; Cambridge and National Institute for Health Research, University of Cambridge, Cambridge, UK*

KITHIGANAHALLI NARAYANASWAMY BALAJI • *Department of Microbiology and Cell Biology, Indian Institute of Science, Bangalore, Karnataka, India*

JAGADEESH BAYRY • *Institut National de la Santé et de la Recherche Médicale Unité 1138, Paris, France; Centre de Recherche des Cordeliers, Unité Mixte de Recherche-Santé 1138, Equipe 16-Immunopathology and Therapeutic Immunointervention, Université Pierre et Marie Curie-Paris 6, Paris, France; Unité Mixte de Recherche-Santé 1138, Université Paris Descartes, Paris, France; INSERM U 1138, Equipe 16-Centre de, Recherche des Cordeliers, Paris, France*

AMIN BOROUJERDI • *Department of Molecular and Experimental Medicine, The Scripps Research Institute, La Jolla, CA, USA*

LESLIE A. BRUGGEMAN • *Department of Medicine and Rammelkamp Center for Education and Research, MetroHealth Medical Center, Case Western Reserve University School of Medicine, Cleveland, OH, USA*

JUAN D. CAÑETE • *Arthritis Unit, Rheumatology Department, Hospital Clinic and IDIBAPS, Barcelona, Spain*

JOSÉ ANTONIO CARMONA ARANA • *Division of Molecular Internal Medicine, Department of Internal Medicine II, University Hospital Würzburg, Würzburg, Germany*

HUI CHEN • *Department of Immunology, Medical College, Wuhan University of Science and Technology, Wuhan, Hubei, China*

HONG R. CHO • *Biomedical Research Center, School of Medicine, Ulsan University Hospital, University of Ulsan, Ulsan, Republic of Korea; Department of Surgery, School of Medicine, Ulsan University Hospital, University of Ulsan, Ulsan, Republic of Korea*

SWATI CHOKSI • *Cell and Cancer Biology Branch, National Cancer Institute, National Institutes of Health, Bethesda, MD, USA*

SRI RAMULU ELLURU • *Singapore Immunology Network, Agency for Science Technology and Research (A*STAR), Singapore, Singapore*

PABLO ENGEL • *Immunology Unit, Department of Cell Biology, Immunology and Neurosciences, Medical School, University of Barcelona and IDIBAPS, Barcelona, Spain*

XU FENG • *Department of Pathology, University of Alabama at Birmingham, Birmingham, AL, USA*

BARBARA FROSSI • *Department of Biological and Medical Science, University of Udine, Udine, Italy*

SAHANA HOLLA • *Department of Microbiology and Cell Biology, Indian Institute of Science, Bangalore, Karnataka, India*

KEIJI IWAMOTO • *Bio-Manufacturing Technology Laboratories, CMC Center, Takeda Pharmaceutical Company Limited, Mitsui, Hikari, Yamaguchi, Japan*

JOEL JULES • *Department of Pathology, University of Alabama at Birmingham, Birmingham, AL, USA*

SRINI V. KAVERI • *Institut National de la Santé et de la Recherche Médicale Unité 1138, Paris, France; Unité Mixte de Recherche-Santé 1138, Equipe 16-Immunopathology and Therapeutic Immunointervention, Centre de Recherche des Cordeliers, Université Pierre et Marie Curie-Paris 6, Paris, France; Unité Mixte de Recherche-Santé 1138, Université Paris Descartes, Paris, France*

HANGEUN KIM • *Department of Internal Medicine, Saint Louis University, St. Louis, MO, USA*

JUYANG KIM • *Biomedical Research Center, School of Medicine, Ulsan University Hospital, University of Ulsan, Ulsan, Republic of Korea*

ANDREI A. KRUGLOV • *German Rheumatism Research Center, Leibniz Institute, Berlin, Germany; Lomonosov Moscow State University, Moscow, Russia*

ANNA A. KUCHMIY • *German Rheumatism Research Center, Leibniz Institute, Berlin, Germany; Engelhardt Institute of Molecular Biology, Russian Academy of Sciences, Moscow, Russia; Flanders Institute for Biotechnology, Ghent University, Ghent, Belgium*

BYUNGSUK KWON • *Biomedical Research Center, School of Medicine, Ulsan University Hospital, University of Ulsan, Ulsan, Republic of Korea; School of Biological Science, University of Ulsan, Ulsan, Republic of Korea*

YAN LENG • *Cancer Center, Tongji Medical College, Tongji Hospital, Huazhong University of Science and Technology, Wuhan, Hubei, China*

ZHUOYA LI • *Department of Immunology, Tongji Medical College, Huazhong University of Science and Technology, Wuhan, Hubei, China*

CHUANJU LIU • *Department of Orthopaedic Surgery, New York University Medical Center, New York, NY, USA; Department of Cell Biology, New York University Medical Center, New York, NY, USA*

ZHENGGANG LIU • *Cell and Cancer Biology Branch, National Cancer Institute, National Institutes of Health, Bethesda, MD, USA*

RICHARD MILNER • *Department of Molecular and Experimental Medicine, The Scripps Research Institute, La Jolla, CA, USA*

SERGEI A. NEDOSPASOV • *German Rheumatism Research Center, Leibniz Institute, Berlin, Germany; Engelhardt Institute of Molecular Biology, Russian Academy of Sciences, Moscow, Russia; Lomonosov Moscow State University, Moscow, Russia*

YELENA L. POBEZINSKAYA • *Cell and Cancer Biology Branch, National Cancer Institute, National Institutes of Health, Bethesda, MD, USA*

CARLO PUCILLO • *Department of Biological and Medical Science, University of Udine, Udine, Italy*

RANJIT RAY • *Departments of Internal Medicine, Molecular Microbiology and Immunology, Saint Louis University, St. Louis, MO, USA*

XAVIER ROMERO • *Immunology Unit, Department of Cell Biology, Immunology and Neurosciences, Medical School, University of Barcelona and IDIBAPS, Barcelona, Spain*

STEFFEN SALZMANN • *Division of Molecular Internal Medicine, Department of Internal Medicine II, University Hospital Würzburg, Würzburg, Germany*

YURY V. SHEBZUKHOV • *German Rheumatism Research Center, Leibniz Institute, Berlin, Germany*

YASUSHI SHINTANI • *Pharmaceutical Research Division, Takeda Pharmaceutical Company Limited, Fujisawa, Kanagawa, Japan*

RICCARDO SIBILANO • *Department of Pathology, Stanford University School of Medicine, Stanford, CA, USA*
VOLKER SIFFRIN • *University Medical Center, Mainz, Germany*
QINGYUN TIAN • *Department of Orthopaedic Surgery, New York University Medical Center, New York, NY, USA*
ULRICH TIGGES • *Department of Molecular and Experimental Medicine, The Scripps Research Institute, La Jolla, CA, USA*
JOHANNES TREBING • *Division of Molecular Internal Medicine, Department of Internal Medicine II, University Hospital Würzburg, Würzburg, Germany*
JAMMA TRINATH • *Department of Microbiology and Cell Biology, Indian Institute of Science, Bangalore, Karnataka, India*
ISAMU TSUJI • *Biomolecular Research Laboratories, Pharmaceutical Research Division, Takeda Pharmaceutical Company Limited, Fujisawa, Kanagawa, Japan*
VOLKER VIELHAUER • *Nephrologisches Zentrum, Medizinische Klinik und Poliklinik IV, Klinikum der Universität München, Ludwig-Maximilians-University Munich, Munich, Germany*
HARALD WAJANT • *Division of Molecular Internal Medicine, Department of Internal Medicine II, University Hospital Würzburg, Würzburg, Germany*
JENNIFER V. WELSER-ALVES • *Department of Molecular and Experimental Medicine, The Scripps Research Institute, La Jolla, CA, USA*
ZHENZHEN WU • *Department of Medicine and Rammelkamp Center for Education and Research, MetroHealth Medical Center, Case Western Reserve University School of Medicine, Cleveland, OH, USA*
BO XIE • *State Key Laboratory of Biochemical Engineering, Institute of Process Engineering, Chinese Academy of Sciences, Beijing, China*
SHUAI ZHAO • *Department of Orthopaedic Surgery, New York University Medical Center, New York, NY, USA*
FRAUKE ZIPP • *University Medical Center, Mainz, Germany*

Chapter 1

Assaying NF-κB Activation and Signaling from TNF Receptors

Zhenzhen Wu and Leslie A. Bruggeman

Abstract

Tumor necrosis factor (TNF)-mediated activation of the NF-κB family of transcription factors is mediated by two receptors, TNFR1 and TNFR2. These two receptors have unique roles in response to TNF and have been the focus of new therapeutic strategies in a variety of diseases with an immune or an inflammatory component. This chapter describes in vitro methods to functionally identify which TNF receptor is initiating NF-κB activation. This will include antibody-mediated receptor blockade and RNAi-mediated gene silencing targeting the individual receptors. The NF-κB activation methods presented include standard, accepted assays for monitoring the sequential activation steps through the NF-κB signal transduction cascade, including IκBα degradation, NF-κB nuclear translocation, and transcriptional activation of gene expression.

Key words TNF receptor, NF-κB, Western blotting, Transient transfection, RNAi, Immunocytochemistry, Transcriptional regulation, Receptor blockade

1 Introduction

The receptors in the tumor necrosis factor (TNF) superfamily that mediate the physiological response to TNF have become important therapeutic targets for the treatment of many chronic inflammatory conditions [1]. TNF can induce multiple intracellular responses including apoptosis and activation of the pro-inflammatory transcription factors NF-κB and AP-1 directly or through downstream ERK, p38MAPK, and JNK signaling [2]. In addition, it is becoming clear that the two major receptors for TNF, TNFR1 and TNFR2, have different roles in pathological processes. This is of particular relevance for TNFR2, which has a more limited tissue distribution but can be induced in disease settings, as compared to TNFR1, a ubiquitous and constitutively expressed receptor. For example, in chronic renal diseases such as HIV-associated nephropathy, glomerulonephritis, and renal allograft rejection, TNFR2 expression is induced on renal epithelial cells and is critical for initiating or exacerbating the underlying pathogenic process [3–5].

Jagadeesh Bayry (ed.), *The TNF Superfamily: Methods and Protocols*, Methods in Molecular Biology, vol. 1155,
DOI 10.1007/978-1-4939-0669-7_1, © Springer Science+Business Media New York 2014

Thus, establishing which TNFR mediates pro-inflammatory processes through NF-κB is useful in establishing a specific, pathogenic event that could be effectively targeted with available therapies.

The NF-κB transcription factor family is central mediators of many immune and inflammatory processes, and TNF is a major physiological stimulus for NF-κB [6]. The two TNF receptors, TNFR1 and TNFR2, have similar extracellular domains, but different intracellular domains that initiate alternative signaling pathways, although both can activate NF-κB (Fig. 1). Upon receptor engagement, the intracellular domain of TNFR1 initially recruits TRADDs and primarily activates caspase-dependent apoptosis but can alternatively recruit TRAF2 and activate NF-κB similar to TNFR2, which binds TRAF2 directly. Both receptor-initiated events converge on the IκBα kinase (IKK) complex. The active IKK complex phosphorylates IκBα, a binding partner of the prototypic NF-κB heterodimer p50/p65, and functions as an inhibitor by anchoring NF-κB in the cytoplasm. Once phosphorylated, IκBα is recognized for ubiquitination and is degraded by the proteosome. The cytoplasmic removal of IκBα permits NF-κB p50/p65 to translocate to the nucleus where it can bind its target gene promoters and activate new gene transcription.

The methods presented here correspond to the sequential events in the NF-κB signal transduction cascade including (1) receptor engagement, (2) release from IκBα and p50/p65 nuclear translocation, and (3) transactivation of gene expression (Fig. 1). We have presented complementary methods in each section which can be used either as alternative strategies or as is frequently needed for peer review and publication, an additional method to confirm observations of the first method.

2 Materials

2.1 Cell Culture

1. Adherent cell line of interest.
2. Culture medium for the cell line used.
3. 24-well tissue culture plates.
4. Hemocytometer or other methods to count cells.
5. Humidified, CO_2, and temperature-controlled incubator.

2.2 TNF Receptor Utilization

1. Recombinant TNF (*see* **Note 1**).
2. TNFR1 blocking IgG (hamster anti-mouse, clone 55R-170, available from many sources; we use BD Biosciences).
3. TNFR2 blocking IgG (hamster anti-mouse, clone TR75-54.7, available from many sources; we use BD Biosciences).
4. Nonspecific hamster IgG (clone HTK888, BD Biosciences).

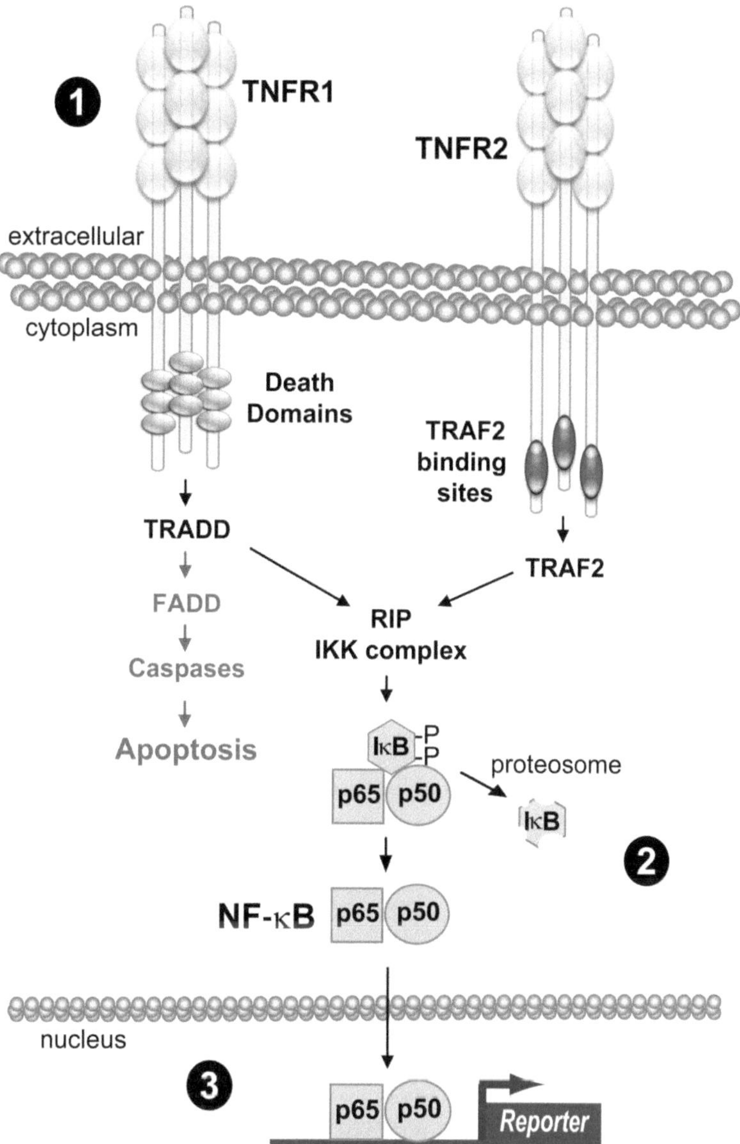

Fig. 1 Generalized diagram of the NF-κB activation pathway induced by TNF receptors. Both TNFR1 and TNFR2 can activate NF-κB signaling, mediated through either TNFR1-associated death domains (TRADD) or TNFR-associated factors (TRAF2), respectively, subsequently converging on the IκB kinase (IKK) complex resulting in the phosphorylation of IκBα and degradation by the proteosome. NF-κB p50/p65, now free of IκBα, translocates to the nucleus and activates new transcription of target gene expression. Key points through the signal transduction cascade to be examined by the assays described here include (*1*) receptor-specific activation, (*2*) IκBα degradation and NF-κB translocation to the nucleus, and (*3*) transactivation of gene expression

5. Pre-made siRNAs for TNFR1, TNFR2, and a non-target control (ON-TARGETplus SMARTpool, Dharmacon).
6. Liposome transfection reagent (Fugene 6, Promega).
7. Rabbit polyclonal antibodies to TNFR1 and TNFR2 (GenTex, Inc).

2.3 Immunofluorescence Localization for p65 Nuclear Translocation

1. #1 German glass cover slips, 9 mm round (sterile).
2. 1 × 3 in. glass slides.
3. Rabbit polyclonal anti-p65/RelA antibody (Santa Cruz Biotechnology, C-19).
4. Fluorescently tagged goat anti-rabbit antibody (Jackson ImmunoResearch Laboratories).
5. Phosphate-buffered saline (PBS): 137 mM NaCl, 2.7 mM KCl, 8 mM Na_2HPO_4, 2 mM KH_2PO_4.
6. Immunofluorescence (IF) block: PBS with 2 % teleostean (fish) gelatin, 2 % BSA, 2 % fetal calf serum.
7. Dulbecco's PBS (DPBS): PBS with 0.5 mM $MgCl_2$, 0.9 mM $CaCl_2$.
8. DPBS-serum buffer: DPBS with 2 % normal goat serum.
9. PBS-Tween buffer: PBS with 0.05 % Tween-20.
10. Fluorescence mounting medium.
11. Epi-fluorescence or confocal microscope.

2.4 Western Blotting for IκBα Degradation

1. Rabbit polyclonal anti-IκBα antibody (Cell Signaling).
2. HRP-tagged goat anti-rabbit secondary antibody (Vector Labs).
3. Protease inhibitors ("Complete" protease inhibitor cocktail, Roche).
4. Assay to determine protein concentration (we use BioRad Protein Assay in the microassay format, BioRad).
5. Chemiluminescence detection solutions for Western blotting.
6. Low detergent block: 50 mM Tris–HCl, pH 8.0, 2 mM $CaCl_2$, 80 mM NaCl, 5 % (w/v) nonfat dry milk, 0.2 % (v/v) NP-40, 0.02 % (w/v) NaN_3.
7. Lysis buffer: 50 mM Tris HCl, pH 8, 150 mM NaCl, 1 % NP-40, 0.1 % SDS, 0.5 % sodium deoxycholate.
8. Tris-glycine SDS running buffer (Life Technologies).
9. Transfer buffer: Tris-glycine native running buffer, 10 % methanol.
10. 4X electrophoresis sample buffer: 200 mM Tris–HCl, pH 6.8, 4 % sodium dodecyl sulfate, 40 % glycerol, 80 mM dithiothreitol, 1.5 mM β-mercaptoethanol, 0.5 mM bromophenol blue.

11. Stripping buffer: Tris–HCl, pH 6.8, 2 % sodium dodecyl sulfate, 100 mM β-mercaptoethanol.
12. PVDF membrane.
13. Microfuge and 1.5 mL microcentrifuge tubes.
14. Precast 4–20 % gradient Tris-glycine denaturing polyacrylamide gel.
15. Protein gel electrophoresis and electroblotting equipment and power supply.
16. Equipment to visualize chemiluminescence for Western blotting (X-ray film or other detectors).
17. Antibodies specific for TNFR1 and TNFR2 (GenTex, Inc, GTX19139 and GTX15563 respectively).

2.5 Transient Transfection and Reporter Assay

1. NF-κB report plasmid suitable for chemiluminescence (pNF-κB-SEAP), positive control reporter pSEAP2-Control, negative control reporter pTAL-SEAP (Clontech).
2. Liposome transfection reagent (Fugene 6).
3. Chemiluminescent SEAP Reporter Gene Assay (Roche).
4. Microfuge and 1.5 mL microcentrifuge tubes.
5. Equipment to visualize chemiluminescence for reporter assays (luminometer or spectrophotometer that can detect direct emitted light, plate reader format) and appropriate 96-well plates (*see* **Note 2**).

3 Methods

3.1 TNF Receptor Utilization by Antibody-Mediated Receptor Blockade

Identifying which TNF receptor is initiating the NF-κB activation cascade can be determined using methods to functionally neutralize the expression of the individual receptors. This can be accomplished by receptor blockade with neutralizing antibodies. Antibody neutralization is achieved through preincubation of cells with receptor-specific antibodies to the extracellular domain that do not have agonistic activity but function to prevent TNF from engaging the receptor [7].

1. The day before the assay, plate cells in standard culture medium in a 24-well dish at a density to reach ~75 % confluence in 0.5 mL of standard growth medium. For reproducibility, it is best to count the cells with a hemocytometer and plate the same number of cells for each experiment. For 24-well dishes, this should be between 10^4 and 10^5 cells/well but will vary depending on cell size and growth rate. If the cells are to be processed for immunofluorescence, plate cells similarly, but place a sterile 9 mm round cover slip into the well before seeding cells.

2. The day of the assay, pretreat cells for 1 h at 37 °C by adding the blocking antibodies or as a negative control a nonspecific (species matched) control antibody, used at a final concentration of 1 μg/mL.
3. While still in the presence of 1 μg/mL of each neutralizing antibody, add 20 ng/mL of TNF directly to the medium. A negative control for the TNF treatment would be the buffer used to dilute the TNF. Incubate cells at 37 °C for either 15 min for Western blotting or 30 min for immunofluorescence (*see* **Note 3**).
4. Determine the effect of receptor blockade on NF-κB activation using either Western blotting to monitor IκBα degradation (*see* Subheading 3.4) or for immunofluorescence to monitor p65/RelA nuclear translocation (*see* Subheading 3.3). If NF-κB activation will be monitored using the functional assay in Subheading 3.5, the blocking antibodies should remain in the medium during the 24-h TNF treatment (*see* alternate **step 4** in Subheading 3.5).

3.2 TNF Receptor Utilization: Gene Silencing by RNAi

Identification of which TNF receptor initiates the NF-κB activation cascade also can be determined by eliminating or significantly reducing receptor expression using RNAi-mediated gene silencing. Silencing or knocking down receptor expression can be achieved through transient transfection of pre-made siRNAs.

1. The day before the assay (at least 16 h in advance of the transfection), plate cells in 0.5 mL of serum-containing medium without antibiotics in 24-well dishes. The cell plating density may depend on your cell line; for example, our renal epithelial cells transfect better at ~75 % confluence, or alternatively, use the plating density recommended by the liposome reagent manufacturer.
2. The day of the assay, transiently transfect pre-made siRNAs for the individual TNFRs or a control siRNA. For each transfection, combine 0.2 μL of Fugene 6 in 20 μL of serum-free basal medium, vortex briefly, and incubate for 5 min at room temperature. Add pre-made siRNAs to 20 pM (2.0 μL of 5 μM solution), vortex briefly, and incubate for 15 min at room temperature. Add the lipid–siRNA complexes to cells, swirl plate to mix, and incubate under normal growth conditions for 48 h.
3. The pre-made siRNAs typically reduce expression by 60–80 %, but as with all siRNA experiments, the degree of silencing should be confirmed by Western blotting. The Western blotting method described in Subheading 3.4 can be used with rabbit polyclonal antibodies for TNFR1 and TNFR2. For this technical control, either plate and treat parallel wells for protein extraction, or alternatively, the cells remaining after the reporter assay (*see* Subheading 3.5) can be harvested for the Western blot.

4. After 48 h, treat transfected wells by adding 20 ng/mL of TNF directly to the medium, swirl to mix, and incubate cells at 37 °C for either 15 min for Western blotting or 30 min for immunofluorescence.
5. Determine the effect of TNFR knockdown on NF-κB activation using immunofluorescence for p65/RelA nuclear translocation (*see* Subheading 3.3) or Western blotting for IκBα degradation (*see* Subheading 3.4). If NF-κB activation will be monitored using the reporter assay in Subheading 3.5, the siRNA transfection should be repeated in a co-transfection with the reporter plasmid to ensure that the knockdown is maintained during the subsequent 48 h of reporter expression (*see* alternate **step 2** in Subheading 3.5).

3.3 NF-κB Activation by Immunofluorescence for p65/RelA Nuclear Translocation

1. At the end of the 30-min incubation period with TNF, place the 24-well dish on ice, and aspirate medium and gently wash twice with ice-cold DPBS. Gentle washing is achieved by touching the pipet tip to the wall of the well and slowly allowing the DPBS to run down the side of the well to flood the cover slip. Repeated application of washes directly onto the cover slip will dislodge even the most adherent cells.
2. Keeping the dish on ice, aspirate final DPBS wash and fix cells by applying 1 ml of −20 °C methanol (*see* **Note 4**) for 5 min. Remove the dish from the ice, and aspirate off the methanol and allow to air-dry completely at room temperature (typically 10–15 min to dry completely).
3. Block for nonspecific antibody binding by applying 0.5 mL of IF block directly to the dry cover slips and incubate at room temperature for 30 min.
4. Remove block, apply 300 μL of the p65/RelA antibody (diluted 1:400 in DPBS-serum buffer), and incubate at room temperature for 1 h. Remove antibody and gently wash three times, 5 min each using DPBS.
5. Remove the final wash, and add 300 μL of the fluorescence-tagged goat anti-rabbit secondary antibody (2 μg/mL in DPBS-serum buffer). Incubate for 1 h at room temperature, protected from light (i.e., cover with aluminum foil). Remove antibody solution and wash three times, 5 min each using DPBS, followed by one wash in ultrapure water.
6. Mount cover slips on a 1 × 3 in. glass slide by placing a 5 μL drop of fluorescent mounting medium on the slide and dropping the inverted (i.e., cell side down) cover slip onto the mounting medium. The cover slips are ready to view by either confocal or epi-fluorescence microscopy or can be stored at 4 °C. An example p65/RelA staining before and after TNF treatment is shown in Fig. 2.

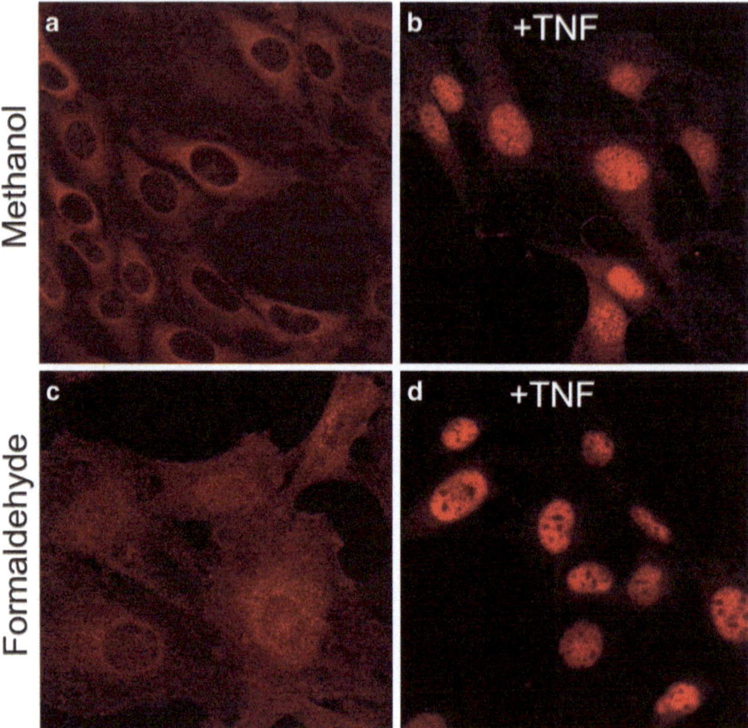

Fig. 2 Immunofluorescence microscopy to visualize subcellular localization of p65/RelA and a fixation artifact. Confocal images of a mouse kidney epithelial (podocyte) cell line treated for 30 min with 20 ng/ml recombinant murine TNF and fixed with either methanol or with neutral buffered 4 % formaldehyde. (**a**) Unstimulated cells with the typical cytoplasmic localization of NF-κB using an antibody to detect p65/RelA. (**b**) Upon TNF treatment, NF-κB translocates to the nucleus. Methanol, a precipitating fixative, is better at preserving the accurate subcellular localization of transcription factors that shuttle between the nucleus and cytoplasm. (**c**, **d**) Formaldehyde, a cross-linking fixative, can cause artifacts that incorrectly indicate a low level of nuclear staining in the unstimulated control (panel **c**) but does produce the correct subcellular localization with TNF treatment (panel **d**)

3.4 NF-κB Activation Western Blotting for IκBα Degradation

1. At the end of the 15-min incubation period with TNF, place the 24-well dish on ice, and aspirate medium and gently wash twice with ice-cold PBS. Be sure to aspirate off all PBS from the final wash (residual PBS will dilute the lysis buffer), but do not let cells dry. Lyse cells in 30 μL of lysis buffer that contains protease inhibitors, and allow the lysis buffer to stand on the wells for 10 min. With a P200 pipette tip, scrape the well and transfer the lysate to a 1.5 mL microcentrifuge tube. This whole-cell lysate will be viscous. Pellet any insoluble material by centrifuging at 10,000 × *g* spin at 4 °C in a microcentrifuge. Transfer the supernatant to a clean tube, and discard the pellet. Samples can be safely stored at −80 °C.

2. Quantify the protein in each sample (*see* **Note 5**), and prepare the samples for electrophoresis by aliquoting an equivalent amount of protein for each sample (20–25 μg for a typical 12–15-well, 1.5 mm thick gel). Add an appropriate volume of 4× sample buffer to make 1× in final concentration, and heat samples at 95 °C for 5 min. Samples can be safely stored at −20 °C.
3. Load and resolve samples on a denaturing 4–20 % gradient polyacrylamide gel. For typical minigel formats, this is at 200 V for 1 h or until the dye front reaches the bottom of the gel.
4. Transfer gel to PVDF membranes using standard electrotransfer procedures using a transfer buffer with 10 % methanol. For typical minigels this is 100 V at 4 °C for 2 h. After transfer, wash the blot for 1 min in 100 % methanol and air-dry. The PVDF membrane now can be safely stored dry at 4 °C or re-wet in 100 % methanol for 1 min, and continue with the Western blotting procedure.
5. Block the wet PVDF membrane with low detergent block for 45 min at room temperature with gentle rocking in an appropriate sized container with sufficient blocking solution to completely cover the membrane.
6. Remove the block, and add the diluted anti-IκBα antibody (1:1,000 dilution in low detergent block solution). Incubate overnight at 4 °C in a container that can be sealed to prevent evaporation.
7. Wash the membrane three times for 10 min each in PBS-Tween buffer at room temperature with gentle shaking.
8. Add the secondary goat anti-mouse HRP antibody (0.1 μg/mL in PBS-Tween buffer) for 1 h at room temperature with gentle shaking.
9. Wash the membrane three times for 10 min each in PBS-Tween buffer at room temperature with gentle shaking.
10. Develop the blots with standard chemiluminescence reagents as recommended by the manufacturer, and expose membrane as required for your method of detection (X-ray film or other detection systems).
11. To confirm equal loading and for normalization, the blot can be stripped and reprobed with an antibody against a control protein that does not change with TNF treatment. Strip the PVDF membranes to remove antibodies by incubating at 50 °C in stripping buffer using a sealed container. Handle the stripping solution in a fume hood due to the strong odor of β-mercaptoethanol. Wash twice in PBS for 10 min each to remove the stripping reagents, and repeat the above procedure from **step 5** substituting the primary antibody in **step 6** with the control antibody (such as α-tubulin). An example of TNF-induced IκBα degradation by Western blotting is shown in Fig. 3a.

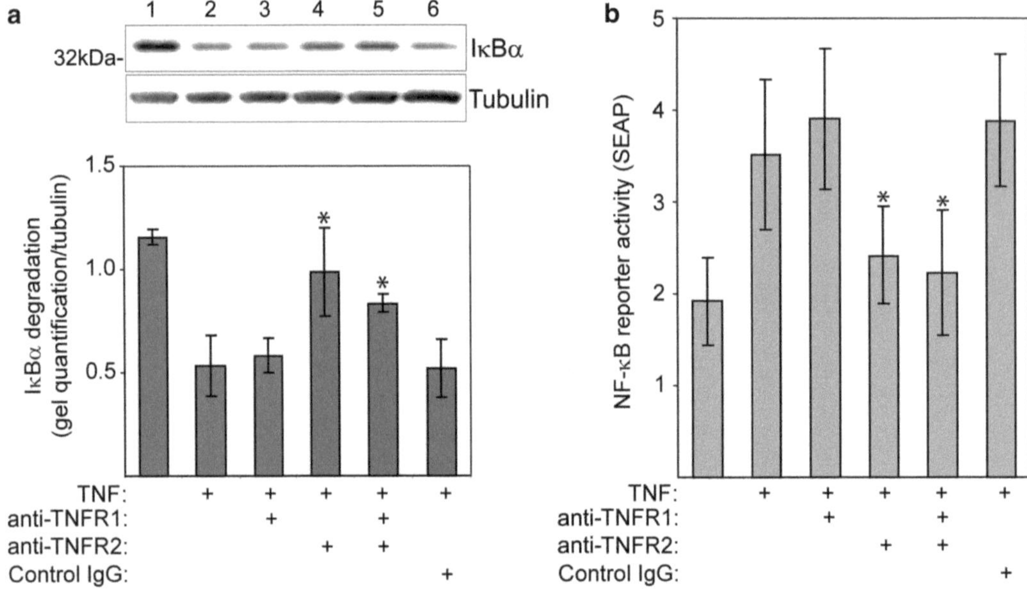

Fig. 3 TNF receptor specificity determined by receptor blockade. A mouse kidney epithelial (podocyte) cell line was pretreated with receptor-blocking antibodies for TNFRI (anti-TNFR1), TNFRII (anti-TNFR2), or control IgG, followed by challenge with TNF. (**a**) Western blotting for IκBα degradation following TNF stimulation demonstrating a block in IκBα degradation with the anti-TNFR2 antibody. A representative Western blot is shown that was re-probed for α-tubulin, a control protein for equal loading, and below is a graph of the quantification of all gels ($n=3$) normalized to α-tubulin expression. *Lanes 1–6* correspond to the treatments shown below the bars on the graph. (**b**) In a similar treatment series as shown in panel **a**, a functional assay for expression of NF-κB-dependent reporter (SEAP) demonstrated a block in NF-κB activation by the anti-TNFR2 treatment. *$P<0.05$ compared to TNF treated. In this experiment TNFR2 blockade, but not TNFR1 blockade, decreased IκBα degradation and expression of the SEAP reporter, and the combined TNFR1 plus TNFR2 blockade was not significantly different than TNFR2 blockade alone, indicating that the TNF-induced NF-κB activation in this cell line was mediated by TNFR2. A version of this image was originally published by the authors in *Laboratory Investigations* [3]

3.5 Functional Assay for NF-κB-Dependent Transcriptional Activation by Reporter Transient Transfection

Transient expression of an NF-κB reporter plasmid can be used as a functional measure of TNF-induced transcriptional activity. For example, an indicator plasmid for NF-κB activity, pNF-κB-SEAP, expresses secreted alkaline phosphatase (SEAP, *see* **Note 6**) under the control of a promoter that contains four NF-κB consensus sites. A positive control plasmid (pSEAP-control) is frequently used to help normalize for differences in transfection efficiency, and a negative control that contains an enhancer/promoter-less SEAP plasmid is also available (pTAL-SEAP) to assess any basal SEAP activity (*see* **Note 6**).

1. The day before the assay (at least 16 h in advance of the transfection), plate cells in 0.5 ml serum-containing medium without antibiotics in 24-well dishes. The cell plating density may depend on your cell line (e.g., our renal epithelial cells transfect better at ~75 % confluence), or alternatively, use the plating

density recommended by the liposome reagent manufacturer. Treatments are usually conducted in duplicate and include additional wells for positive and negative controls.

2. Transiently transfect cells using FuGENE 6 at a 3:1 lipid-to-DNA ratio or according to the manufacturer's recommendations if using a different liposome reagent. For each transfection, combine 0.6 μL of Fugene 6 in 20 μL of serum-free basal medium, vortex briefly, and incubate for 5 min at room temperature. Add 0.2 μg of SEAP plasmid DNA, vortex briefly, and incubate for 15 min at room temperature. Add the lipid–DNA complexes to cells, swirl plate to mix, and incubate under normal growth conditions for 48 h.

   [Alternate **step 2** if using siRNA knockdowns. To maintain siRNA knockdown during reporter expression, transfect cells again with siRNA combining the siRNA and expression plasmid DNA together. In 20 μl of serum-free basal medium add 0.8 μL of Fugene 6, vortex briefly to mix, and incubate for 5 min at room temperature. Add 0.2 μg of SEAP plasmid DNA and 20pM of siRNA, vortex briefly, and incubate lipid–DNA/siRNA complexes at room temperature for 15 min.]

3. Add lipid–DNA complexes directly to cells without changing the medium. Swirl to mix, and incubate for 24 h under normal growth conditions for your cell line.

4. After 24 h, remove the medium and replace with fresh antibiotic-free medium containing 20 ng/mL of TNF. Incubate cells for an additional 24 h under normal growth conditions.

   [Alternate **step 4** if using antibody-mediated receptor blockade. To maintain receptor blockade during reporter expression, remove the medium and replace with fresh antibiotic-free medium that contains 1 μg/ml of the blocking or the control antibodies and incubate for 1 h at 37 °C. After 1 h, do not change medium, but add 20 ng/mL of TNF, and swirl plate to mix. Incubate cells for an additional 24 h under normal growth conditions.]

5. After a total of 48 h post-transfection and 24 h after TNF treatment, sample medium by transferring 0.1 mL of medium to a 1.5 mL microcentrifuge tube. Centrifuge briefly (1 min at 10,000 × *g*) to remove any detached cells or cell debris, and transfer the supernatant to a clean tube. For the SEAP assay, these samples can be safely stored at −20 °C.

6. SEAP activity is assayed from conditioned medium using a chemiluminescence conversion kit as directed by the manufacturer. Chemiluminescence is measured using a luminometer, and the generated number data will be without units (sometimes referred to as relative light units). To normalize for transfection efficiency (1) average duplicates for each sample and

(2) subtract the negative control value (baseline SEAP levels) from each, and (3) to normalize for transfection efficiency for inter-experiment comparisons, the sample values are divided by the positive control value for each experiment. An example of TNF-induced NF-κB-dependent gene expression and inhibition with receptor blockade is shown in Fig. 3b.

7. If the reporter assay was used with a siRNA experiment, these cells can now be harvested for protein extraction and Western blotting to confirm the degree of receptor knockdown.

4 Notes

1. Human and mouse TNF are not interchangeable. It is known that human TNF is not recognized by mouse TNFR2 [8]. Therefore, the species origin of the cell line and the TNF protein should be the same, that is, human TNF should be used in experiment with human cells, and murine TNF should be used in experiments with murine cells. The source of the recombinant protein (i.e., bacteria derived versus mammalian cell derived) does not significantly change functionality, and either can be used. If using bacteria-derived protein, use a source where the quality control specifications include an analysis for endotoxin contamination (should be <1.0 EU per 1 μg protein by the LAL method), since bacterial endotoxins can activate NF-κB.
2. The type of 96-well plate used for reading the chemiluminescence signal will depend on the type of plate reader available. Some machines insert a tube into each well to sample a small volume from each well, and the light is quantified within the instrument. For these types of "sipper" machines any 96-well plate can be used. Other machines directly read the emitted light from the plate. For these machines, the type of 96-well plate is important and should be specified in the user manual for the instrument. Light detectors that read from the top of the plate ("top read") typically require fully opaque (black or white) plates, whereas detectors that read from the bottom of the plate ("bottom read") require opaque plates with clear, flat bottoms.
3. Timing of cell fixation or harvests after TNF treatment is critical. The NF-κB signal transduction cascade occurs in minutes, and unphosphorylated IκBα is quickly recreated as part of the deactivation process [9, 10]. Allowing experiments to proceed longer than 30 min after TNF treatment could allow the cell to return to an unstimulated state, and activation events could be missed. Degradation of IκBα occurs within 1–5 min of TNF treatment, and assay time points are usually within 15 min of TNF treatment. The translocation of Rel/p65 happens subsequent to IκBα degradation, and persists in the nucleus for a

longer period, and assay time points are usually within 30 min. In addition, it is imperative for inter-experiment reproducibility to be consistent with timing from the start of stimulation to termination. This is also important for intra-experiment consistency when doing a large number of samples, where it might not be possible to process all samples quickly enough to maintain a consistent time frame from sample to sample. For experiments with a large number of samples, use of multichannel pipettors would facilitate the synchronous application of fixatives or lysis buffers to uniformly stop the experiment.

4. Cross-linking fixatives such as formaldehyde can cause artifacts with the subcellular localization of p65 and p50, typically producing false nuclear staining (Fig. 2). Precipitating fixatives such as alcohols (methanol) or acetone are superior to cross-linking fixatives such as paraformaldehyde, formaldehyde, formalin, or glutaraldehyde. Other combination fixatives that include both cross-linking and precipitating agents are better than cross-linking agents alone for avoiding this artifact.
5. If the initial cell plating was done accurately and each well contains the same number of cells, the protein determination step can be eliminated as the whole-cell lysates frequently have nearly identical protein concentrations. In this case, the cells can be lysed directly in 25 μL of 1× electrophoresis sample buffer and the gel can be run immediately.
6. An advantage of using the SEAP (or MetLuc) reporter is that the cells are not destroyed by the assay. Thus, after sampling the medium for the SEAP assay, the cell monolayers can be harvested and used for additional assays. For example, cells could be processed for Western blotting or RT/PCR to determine the effect of TNF stimulation on the expression of endogenous NF-κB-regulated genes.

References

1. Croft M, Benedict CA, Ware CF (2013) Clinical targeting of the TNF and TNFR superfamilies. Nat Rev Drug Discov 12:147–168
2. Aggarwal BB, Gupta SC, Kim JH (2012) Historical perspectives on tumor necrosis factor and its superfamily: 25 years later, a golden journey. Blood 119:651–665
3. Bruggeman LA, Drawz PE, Kahoud N, Lin K, Barisoni L, Nelson PJ (2011) TNFR2 interposes the proliferative and NF-kappaB-mediated inflammatory response by podocytes to TNF-alpha. Lab Invest 91:413–425
4. Vielhauer V, Stavrakis G, Mayadas TN (2005) Renal cell-expressed TNF receptor 2, not receptor 1, is essential for the development of glomerulonephritis. J Clin Invest 115: 1199–1209
5. Hoffmann U, Bergler T, Rihm M et al (2009) Upregulation of TNF receptor type 2 in human and experimental renal allograft rejection. Am J Transplant 9:675–686
6. Vallabhapurapu S, Karin M (2009) Regulation and function of NF-kappaB transcription factors in the immune system. Annu Rev Immunol 27:693–733
7. Sheehan KC, Pinckard JK, Arthur CD, Dehner LP, Goeddel DV, Schreiber RD (1995) Monoclonal antibodies specific for murine p55

and p75 tumor necrosis factor receptors: identification of a novel in vivo role for p75. J Exp Med 181:607–617

8. Lewis M, Tartaglia LA, Lee A, Bennett GL, Rice GC, Wong GH et al (1991) Cloning and expression of cDNAs for two distinct murine tumor necrosis factor receptors demonstrate one receptor is species specific. Proc Natl Acad Sci U S A 88:2830–2834
9. Ghosh S, Karin M (2002) Missing pieces in the NF-kappaB puzzle. Cell 109:S81–S96
10. Rice NR, Ernst MK (1993) In vivo control of NF-kappa B activation by I kappa B alpha. EMBO J 12:4685–4695

Chapter 2

Dissecting DR3 Signaling

Yelena L. Pobezinskaya, Zhenggang Liu, and Swati Choksi

Abstract

Receptor signaling can be evaluated in multiple ways, including analysis of phosphorylation of downstream molecules and analysis of proteins that are recruited to the receptor upon ligand binding. Majority of studies on the mechanism of DR3 signaling were performed using overexpression systems that can often lead to artifacts. In this chapter we describe how to analyze DR3 downstream events with most attention being paid to endogenous immunoprecipitation.

Key words Death receptor 3, MAP kinases, NF-kB, Proliferation, T cells, Immunoprecipitation, CFSE

1 Introduction

Death receptor 3 (DR3/TNFRSF25) belongs to the TNF-receptor (TNFR) superfamily and is one of the eight members of the subfamily of death receptors [1]. DR3 is expressed in thymus, spleen, small intestine, and peripheral blood lymphocytes (PBL) and is upregulated in activated T cells [2–6]. The specific ligand for DR3 belongs to the TNF superfamily and is called TNF-like ligand 1A (TL1A/TNFSF15). TL1A has been shown to play an activating role in augmenting T cell responses during various inflammatory and autoimmune conditions including graft-versus-host disease [7], inflammatory bowel disease [8, 9], allergic lung inflammation [10], rheumatoid arthritis, and experimental autoimmune encephalomyelitis [11, 12]. Upon stimulation with TL1A, DR3 like all the other members of the TNFR superfamily forms a signaling complex. TRADD is the primary adaptor protein that organizes the signaling complex and recruits a number of important molecules including RIP1 and TRAF2 that in turn lead to NF-κB and MAPK activation [3, 4]. Despite the fact that DR3 is a death receptor, its ligation does not cause cell death. Instead it leads to the proliferation of primary T cells [13, 14]. Deficiency in TRADD has severe consequences on DR3 downstream signaling.

Jagadeesh Bayry (ed.), *The TNF Superfamily: Methods and Protocols*, Methods in Molecular Biology, vol. 1155, DOI 10.1007/978-1-4939-0669-7_2, © Springer Science+Business Media New York 2014

Receptor signaling complex formation is impaired, and NF-κB and MAPK activation as well as T cell proliferation are reduced in TRADD-deficient cells [14].

In this chapter we describe how to analyze the signaling mediated through DR3. Activation of NF-κB and MAPK can be studied by determining the phosphorylation status of IkBα and MAP kinases using a regular Western blot analysis. The actual procedure of immunoblotting will not be described here; however, we provide all the necessary details on how to treat the cells, how to prepare the samples for Western blot, and what antibodies to use. We focus on two methods: CFSE labeling of T cells and the immunoprecipitation (IP) of endogenous DR3. By utilizing the technique of IP the proteins that are recruited to the receptor-signaling complex can be analyzed. CFSE is a fluorescent dye that enters the cell, and once it is stably incorporated within the cell it is diluted in half with each cell division allowing to follow proliferating cells for up to 6–7 divisions. This process can be visualized by flow cytometry as a series of peaks of decreasing fluorescence intensity.

2 Materials

2.1 Common Materials

1. PBS: 137 mM NaCl, 2.7 mM KCl, 8 mM Na2HPO4, 2 mM KH2PO4.
2. Complete growth medium for primary T cells: RPMI-1640, 2 mM L-glutamine, 100 U/mL penicillin, 100 μg/mL streptomycin, 5×10^{-5} M β-mercaptoethanol, 100 mM HEPES buffer, 10 % fetal bovine serum.
3. Recombinant mouse TL1A, carrier-free (R&D Systems): Reconstitute and store according to the manufacturer's instructions.
4. Anti-CD3ε antibodies for T cell activation (Pharmingen).

2.2 T Cell Purification

1. CD8α^{+} T Cell Isolation Kit II, mouse (Miltenyi Biotec).
2. CD4^{+} T Cell Isolation Kit II, mouse (Miltenyi Biotec).
3. Pan T Cell Isolation Kit II, mouse (Miltenyi Biotec).
4. MACS LS separation columns (Miltenyi Biotec).
5. MACS magnetic stand and magnet (Miltenyi Biotec).

2.3 IP and Immunoblotting

1. Anti-DR3 antibodies (R&D Systems).
2. Antibodies for Western blot analysis: Rabbit anti-pJNK polyclonal antibodies (Life Technologies, Biosource), mouse anti-JNK antibodies (Pharmingen), rabbit anti-pIKBα monoclonal (Cell Signaling).

3. Lysis buffer: M2 buffer (20 mM Tris–HCl, pH 7.5, 0.5 % NP-40, 250 mM NaCl, 3 mM EDTA, 3 mM EGTA) supplemented with 2 mM DTT, 0.5 mM PMSF, 20 mM β-glycerol phosphate, 1 mM sodium vanadate, 1 μg/mL leupeptin (*see* **Note 1**).
4. Protein G-agarose (Roche Applied Science).
5. Microtube rotator.

2.4 CFSE Labeling

1. CFSE labeling kit: VybrantR CFDA SE Cell Tracer Kit (Life Technologies).
2. Water bath.
3. FACSCalibur (BD Biosciences).

3 Methods

3.1 Activation of T Cells

Before treatment with TL1A, purified primary T cells (*see* **Note 2**) should be activated with anti-CD3 antibodies.

1. Day 1: Coat the 10 cm dish with 5 mL of 1 μg/mL of anti-CD3 antibodies overnight at 4 °C (*see* **Note 3**).
2. Day 2: Wash the dish with PBS twice, and plate purified T cells (*see* **Note 4**) in complete growth medium overnight.
3. Day 3: Collect the cells, and wash them once with PBS. Resuspend the cells in fresh medium, and count the cells. T cells are now ready for the experiments.

3.2 Preparation of Samples for Western Blotting

1. Collect plate-activated T cells (*see* **Note 5**), and treat them with 50 μg/mL of TL1A for different time depending on your experiment (*see* **Note 6**).
2. Prepare 1.5-mL microcentrifuge tubes on ice.
3. Collect the cells from each dish or well to a separate 1.5-mL microcentrifuge tube.
4. Centrifuge tubes for 5 min at 300 × *g* at 4 °C.
5. Aspirate the supernatant, add 1 mL of ice-cold PBS, and centrifuge for 5 min at 300 × *g* at 4 °C.
6. Aspirate supernatant, and resuspend the pellet in 30 μL of M2 lysis buffer.
7. Rotate tubes in a rotator for 30 min at 4 °C.
8. Centrifuge tubes for 10 min at 15,000–20,000 × *g* at 4 °C.
9. Transfer the supernatant to a fresh 1.5-mL microcentrifuge tube without disturbing the pellet. The supernatant can be frozen at −70 °C.
10. Measure the concentration of proteins in the sample.
11. Analyze by immunoblotting of your choice.

3.3 CFSE Labeling

1. Collect activated T cells into 15 or 50 mL tube, wash twice with PBS, and transfer to a new tube before the final spin.
2. Resuspend the cells at 1×10^6 per mL in pre-warmed PBS and incubate in 37 °C water bath for at least 10 min. If you have less than 1×10^6 cells still resuspend in 1 mL of pre-warmed PBS.
3. Briefly spin aliquot of 5 mM stock solution of CFSE and dilute 1:3 in PBS (for example, add 10 μL of CFSE into 20 μL of PBS) (*see* **Note 7**).
4. Add 1 μL of diluted CFSE per 1 mL of warmed cells. Add to the side of the tube, and mix immediately by inversion.
5. Incubate in 37 °C water bath for 10 min mixing every few minutes.
6. Stop reaction with ice-cold complete RPMI 1640 medium (at least 10 mL) and centrifuge for 5 min at $300 \times g$.
7. Resuspend in fresh medium, count the cells and plate in 24-well plates at 0.5×10^6/well, and treat with 50 μg/mL of TL1A according to your experiment.
8. Collect cells in 72 h, stain them with antibodies if necessary, and analyze by flow cytometry (*see* **Note 8**).

3.4 Immuno-precipitation

1. Preparation of the cells: Plate 7–8×10^7 activated (*see* Subheading 3.1) T cells in a 15-cm dish and treat with 50 μg/mL of TL1A for 5 min (*see* **Note 9**).
2. Preparation of the cell lysate:
   (a) Prepare 50 mL tube with 20 mL of ice-cold PBS.
   (b) After treatment with TL1A, quickly transfer the cells from the plate to the tube with PBS (*see* **Note 10**). Collect the residual cells with an additional 10 mL of ice-cold PBS.
   (c) Centrifuge tubes for 5 min at $300 \times g$ at 4 °C.
   (d) Aspirate supernatant, and resuspend the pellet in 1 mL of M2 lysis buffer. The pellet can be frozen at −70 °C.
   (e) Transfer the supernatant to a 1.5-mL microcentrifuge tube.
   (f) Rotate on a rotator for 30 min at 4 °C.
   (g) Centrifuge tubes for 10 min at 15,000–20,000 g at 4 °C.
   (h) Transfer the supernatant to a fresh 1.5-mL microcentrifuge tube without disturbing the pellet. The supernatant can be frozen at −70 °C.
   (i) Measure the concentration of proteins in the sample. Take 1 mg for IP and 1–3 % (w/w) for the input (loading control).

3. Washing the beads and preparation of 50 % bead slurry:
   (a) Take enough quantity of protein G-agarose beads (*see* **Note 11**).
   (b) Centrifuge for 1 min at 500 × *g* at 4 °C.
   (c) Resuspend the pellet in the lysis buffer equal to the volume of beads taken. This is now 50 % bead slurry.
4. Preclearance of the lysate:
   (a) Bring the volume of the protein lysate taken for IP to 1 mL with the lysis buffer and add 40 μL of 50 % bead slurry.
   (b) Rotate on a rotator for 1 h at 4 °C.
   (c) Centrifuge for 2 min at 500 × *g* at 4 °C.
   (d) Transfer the supernatant to a fresh 1.5-mL microcentrifuge tube. Be careful not to transfer beads as the bead pellet is very loose.
5. Immunoprecipitation:
   (a) Add 1 μg of DR3 antibodies and 30 μL of 50 % protein G-agarose bead slurry to 1 mL of cell lysate.
   (b) Rotate tubes on a rotator overnight at 4 °C (*see* **Note 12**).
   (c) Centrifuge tubes for 2 min at 500 × *g* at 4 °C.
   (d) Very carefully aspirate the supernatant leaving some liquid above the pellet and add 1 mL of M2 lysis buffer (*see* **Note 13**).
   (e) Rotate tubes on a rotator for 10 min at 4 °C.
   (f) Centrifuge tubes for 2 min at 500 × *g* at 4 °C.
   (g) Wash beads four more times (repeat **step 4d–f**).
   (h) After the final wash, aspirate as much supernatant as possible, and resuspend the beads in 40 μL of loading buffer.
   (i) Denature immunoprecipitate samples and input samples for 1 min at 100 °C.
   (j) Analyze by immunoblotting of your choice (*see* **Note 14**).

4 Notes

1. Basic M2 buffer can be prepared in a large amount and stored at 4 °C. If protease inhibitors are added lysis buffer should be stored in aliquots at −20 °C. After thawing discard unused buffer.
2. Both $CD4^+$ and $CD8^+$ T cells can be used to study DR3 signaling. Purification of lymph node T cells can be performed by using MACS T cell isolation kits (Miltenyi Biotec) according to the manufacturer's instructions. If $CD4^+$ or $CD8^+$ T cell

(cytotoxic T lymphocyte) separation is not needed, Pan T cell isolation kit can be used instead. Keep in mind that you will lose cells after purification on the column. You will get approximately five times less cells than the starting amount. It is possible to get up to 6×10^7 cells from lymph nodes of one mouse.

3. Original vial of anti-CD3ε antibodies is 1 mg/mL. Make sure that it is NA/LE (no-azide/low endotoxin). Working solution of 1 μg/mL in PBS should be used immediately and cannot be reused.
4. You can plate up to 7×10^7 cells per 10 cm dish in 10 mL of media. Just make sure that all the cells touch the bottom of the plate and are in contact with anti-CD3ε antibodies.
5. 2×10^6 T cells plated in 35 mm dish or in one well of 6-well plate is enough for one Western blot. If you think of running more than one gel increase the amount of cells. In theory one mouse should be enough to analyze signaling events by Western blot (*see* **Note 2**). But to be safe start with two mice.
6. When looking for MAP kinase or NFκB activation treatments are usually very short. It is always good to do a time course of 5, 15, 30, and 60 min. There are two options of treating and collecting cells. In the first option all the treatments can be done simultaneously and cells can then be collected at different times. In this case cells should be plated in 35 or 60 mm dishes so that when you for example take your 5-min time point other cells are not disturbed. Alternatively cells can be treated at different times starting with the longest one and then collected simultaneously. In this case cells should be plated in a 6-well plate and treatments should be done without taking the cells out of the incubator again for the reason of not disturbing and stressing out the cells every time you treat. When collecting the cells try to be as fast as possible. Do not let 5 min transform into 10 or 15 min. It is especially important when you are comparing cells from different mice.
7. Prepare the stock solution of CFSE according to the manufacturer's instructions. Always use as less CFSE as possible because it may be toxic to cells. We found that the lowest working concentration suggested in the protocol (0.5 μM) worked best for T cells. The concentration of the stock solution was reduced to 5 mM as opposed to suggested 10 mM. Aliquot the stock and freeze at −20 °C. Do not freeze–thaw, and discard unused reagent.
8. Cells labeled with CFSE can be detected in the FL1 channel, which means that FITC-labeled antibodies cannot be used for staining.
9. $7–8 \times 10^7$ T cells is enough to yield around 1 mg of the protein, which is the amount used for each IP. You will need minimum two samples (untreated and treated) if doing one time point.

As was mentioned in **Note 2** you will get approximately five times less cells than the starting amount after purification on the column. Thus you will need to start with 70×10^8 total lymph node cells, which is about ten mice.

10. Signaling complexes are formed within 5 min of treatment with the ligand. It is important to keep everything on ice and lyse the cells quickly after treatment. Cells may be collected while 5 min is still going.
11. For each sample you will need 40 μL of beads. Since the beads will be used twice (to preclear the lysate and for the actual IP) the total amount should be doubled.
12. Sometimes when we probed the Western blot of IP samples with anti-TRADD antibodies we saw a nonspecific band in untreated lanes and even in TRADD KO lanes. Three things can be done to decrease nonspecific binding. First, cell lysates can be incubated overnight with DR3 antibodies only and agarose beads can be added the next day for 5–6 h. Second, cell lysates can be incubated overnight with the beads only and DR3 antibodies can be added the next day for several hours. It is recommended to try both ways and determine what works better. Third, number and duration of washes can be increased.
13. The bead pellet is very loose. Be very careful. We usually use protein gel-loading tips or 30 G needles to suck the supernatant off.
14. Antibodies used to detect proteins in the signaling complex are anti-RIP 1 antibodies (BD Transduction Laboratories), anti-TRADD antibodies, and anti-TRAF2 antibodies (both from Santa Cruz Biotechnology). When probing with RIP1 antibodies in addition to the actual RIP1 band you should be able to see high-molecular-weight smear which corresponds to ubiquitination of RIP.

References

1. Lavrik I, Golks A, Krammer PH (2005) Death receptor signaling. J Cell Sci 118:265–267
2. Bodmer JL, Burns K, Schneider P, Hofmann K, Steiner V, Thome M et al (1997) TRAMP, a novel apoptosis-mediating receptor with sequence homology to tumor necrosis factor receptor 1 and Fas(Apo-1/CD95). Immunity 6:79–88
3. Chinnaiyan AM, O'Rourke K, Yu GL, Lyons RH, Garg M, Duan DR et al (1996) Signal transduction by DR3, a death domain-containing receptor related to TNFR-1 and CD95. Science 274:990–992
4. Kitson J, Raven T, Jiang YP, Goeddel DV, Giles KM, Pun KT et al (1996) A death-domain-containing receptor that mediates apoptosis. Nature 384:372–375
5. Screaton GR, Xu XN, Olsen AL, Cowper AE, Tan R, McMichael AJ et al (1997) LARD: a new lymphoid-specific death domain containing receptor regulated by alternative pre-mRNA splicing. Proc Natl Acad Sci U S A 94:4615–4619
6. Tan KB, Harrop J, Reddy M, Young P, Terrett J, Emery J et al (1997) Characterization of a novel TNF-like ligand and recently described

TNF ligand and TNF receptor superfamily genes and their constitutive and inducible expression in hematopoietic and non-hematopoietic cells. Gene 204:35–46

7. Migone TS, Zhang J, Luo X, Zhuang L, Chen C, Hu B et al (2002) TL1A is a TNF-like ligand for DR3 and TR6/DcR3 and functions as a T cell costimulator. Immunity 16: 479–492
8. Bamias G, Martin C 3rd, Marini M, Hoang S, Mishina M, Ross WG et al (2003) Expression, localization, and functional activity of TL1A, a novel Th1-polarizing cytokine in inflammatory bowel disease. J Immunol 171:4868–4874
9. Bamias G, Mishina M, Nyce M, Ross WG, Kollias G, Rivera-Nieves J et al (2006) Role of TL1A and its receptor DR3 in two models of chronic murine ileitis. Proc Natl Acad Sci U S A 103:8441–8446
10. Fang L, Adkins B, Deyev V, Podack ER (2008) Essential role of TNF receptor superfamily 25 (TNFRSF25) in the development of allergic lung inflammation. J Exp Med 205:1037–1048
11. Meylan F, Davidson TS, Kahle E, Kinder M, Acharya K, Jankovic D et al (2008) The TNF-family receptor DR3 is essential for diverse T cell-mediated inflammatory diseases. Immunity 29:79–89
12. Pappu BP, Borodovsky A, Zheng TS, Yang X, Wu P, Dong X et al (2008) TL1A-DR3 interaction regulates Th17 cell function and Th17-mediated autoimmune disease. J Exp Med 205:1049–1062
13. Chen NJ, Chio II, Lin WJ, Duncan G, Chau H, Katz D et al (2008) Beyond tumor necrosis factor receptor: TRADD signaling in toll-like receptors. Proc Natl Acad Sci U S A 105: 12429–12434
14. Pobezinskaya YL, Choksi S, Morgan MJ, Cao X, Liu ZG (2011) The adaptor protein TRADD is essential for TNF-like ligand 1A/ death receptor 3 signaling. J Immunol 186: 5212–5216

Chapter 3

Modulation of FcεRI-Dependent Mast Cell Response by OX40L

Riccardo Sibilano, Carlo Pucillo, and Barbara Frossi

Abstract

OX40L is expressed by many cell types, including antigen presenting cells (APCs), T cells, vascular endothelial cells, mast cells (MCs), and natural killer cells. The importance of OX40L:OX40 interactions and the OX40L signaling is crucial for the homeostasis and for the modulation of the effector functions of the immune system. However, the lack of non-murine/non-IgG commercially available OX40L-triggering antibodies and the potential signal cross-contamination caused by the binding to the FcγRs co-expressed by several immune cells have limited the study of the OX40L-signaling cascade. We recently characterized the functions and described the molecular events, which follow the engagement of OX40L in MCs, by the use of the soluble OX40 molecule, able to mimic the regulatory T cell-driven engagement of MC-OX40L. This molecule enables signaling studies in MCs with any requirement for OX40-expressing cells. Using this unique reagent, we determined the modality and the extent by which the engagement of OX40L in MCs influences the IgE-dependent MC degranulation. This tool may find a potential application for signaling studies of other OX40L-expressing populations other than MCs, mainly APCs, with similar approaches we reported for the study of OX40L cascade.

Key words OX40L, Mast cells, OX40, T cells, IgE, Degranulation

1 Introduction

MCs are considered the major effectors of allergy and immediate hypersensitivity reactions [1]. The antigen (Ag)-dependent aggregation of high-affinity IgE receptors (FcεRIs) mediates the activation of MCs and the release of a wide range of proinflammatory compounds such as histamine, leukotrienes, and cytokines [2].

Recent evidences show that MCs can selectively regulate the extent of their activation and orchestrate the immune inflammation by physically interacting, via a series of membrane-bound co-receptors, with other immunological partners, such as eosinophils, B cells, effector and regulatory T cells (Tregs) [3]. The engagement of adhesion molecules (ICAM, CD226), members of the B7 family (ICOSL, PD-L1, and PD-L2), and members of the tumor

Jagadeesh Bayry (ed.), *The TNF Superfamily: Methods and Protocols*, Methods in Molecular Biology, vol. 1155, DOI 10.1007/978-1-4939-0669-7_3, necrosis factor (TNF)/TNF receptor (TNFR) superfamily (OX40L, CD153, Fas, 4-1BB, and GITR) on MC membrane by corresponding ligands modulates MC activation [3]. In particular, we and others reported the presence and studied the function of the OX40L molecule on MCs [3–5]. OX40L (known as CD252, TNFSF4) was first identified as gp34 protein on HTLV-I-transformed cells [6] and later characterized as the ligand for OX40 [7]. The constitutive expression of OX40L on both mouse and human MCs suggests a common pathway of interaction with many OX40-expressing partners [8]. Among the entire T cell repertoire, only natural regulatory T cells (Tregs) constitutively express OX40 on their membrane [9]. We recently identified the mechanism by which the engagement of OX40L on MC membrane by OX40-expressing Tregs resulted in the reduction of Ag-dependent degranulation [10]. In vivo loss of OX40L:OX40 axis functionality resulted in the impairment of MC:Treg cross-talk and caused deregulation of IgE-dependent responses [10, 11].

OX40L triggering in Ag-activated MCs causes a defined signaling cascade, with the recruitment of the C-terminal Src kinase to lipid rafts, the inactivation of Fyn kinase, PI3K, Akt and RhoA, events that impair microtubule formation and eventually, markedly reduced the exocytosis of granule-stored MC mediators [12].

To identify the molecular events following OX40L triggering, we developed [13] the soluble form of the OX40 molecule (sOX40) that selectively triggered MCs without the physical requirement of Tregs. The development of the sOX40 has been necessary in order to overcome several limitations related to the use of OX40L triggering antibodies for signaling studies: (1) the few commercially available IgG-OX40L agonist antibodies might bind Fcγ receptors expressed on MCs and might deliver additional signals which would impair the identification of OX40L effect on FcεRI activation [14]; (2) supernatants of cells transfected with plasmids encoding for the OX40 molecule may cause possible cross-contaminations due to the lack of purity of the secreted products; (3) the use of Tregs to stimulate OX40L on MC surface is limited by the inability to uniquely dissect the OX40L-dependent signal from other potential interaction(s) between MCs and Tregs. Moreover, the OX40L:OX40 cross-talk is altered in many pathological conditions such as IgE-mediated responses, autoimmune diseases, and several cancer models [10, 15] and the use of the sOX40 in such settings may represent a concrete possibility for in vivo studies.

Here we report a detailed protocol to obtain the sOX40 from the OX40-IgG1 fusion protein. The Fc portion of the protein is removed by enzymatic digestion and the sOX40 is recovered and is used to aggregate OX40L during the challenge of IgE-presensitized MCs with the specific Ag.

2 Materials

2.1 Reagents for Differentiation and Activation of Mouse Bone Marrow-Derived MCs (BMMCs)

1. 4–8 week old C57BL/6 mice.
2. Recombinant mouse IL-3 (PeproTech, UK).
3. Mouse anti-FcεRI antibody (clone MAR-1, eBioscence, USA).
4. Mouse DNP-specific IgE.
5. DNP-human serum albumin (DNP_{36}-HSA, Sigma-Aldrich, USA).
6. p-NAG (N-acetyl-*beta*-glucosaminidase, Sigma-Aldrich, USA).
7. IL-3 (PeproTech).
8. Anti-mouse FcεRI antibody (eBioscience).

2.2 Reagents Required for sOX40 Purification and Analysis

1. OX40-IgG1 fusion protein containing the mouse OX40 extracellular domain (Genentech, USA).
2. Papain, powder purified from *Papaya latex* (Sigma-Aldrich, USA).
3. Iodoacetamide (Sigma-Aldrich).
4. Protein G Sepharose 4 fast Flow (GE Healthcare, USA).

2.3 Buffers

1. BMMC culture medium: RPMI 1640 medium, 20 % fetal bovine serum (FBS), 100 U/ml penicillin, 100 μg/ml Streptomycin, 2 mM glutamine, 20 mM Hepes, 1× nonessential amino acids (from 100× mix), 1 mM Sodium Pyruvate (all from Euroclone, Italy), 50 mM β-mercaptoethanol (Sigma-Aldrich).
2. Low-percentage FBS, cytokine-free, BMMC culture medium: BMMC culture medium with 5 % FBS.
3. Phosphate buffer saline (PBS), pH 7.4: 137 mM NaCl, 2.7 mM KCl, 8 mM Na_2HPO_4, 2 mM KH_2PO_4.
4. Tyrode's Buffer: 10 mM Hepes–NaOH, pH 7.4, 130 mM NaCl, 5 mM KCl, 1.4 mM $CaCl_2$, 1 mM $MgCl_2$, 5.6 mM glu cose, 0.1 % BSA.
5. Lysis Buffer: Tyrodes Buffer, 0.5 % Triton-X 100.
6. sOX40 digestion buffer: 0.1 M phosphate buffer–NaOH, pH 7, 10 mM cysteine, 2 mM EDTA.
7. p-NAG buffer: 0.1 M citrate buffer–HCl, pH 4.5, 1 mM p-NAG.
8. Carbonate buffer: 0.1 M $Na_2CO_3/NaHCO_3$.
9. Phosphate buffer (0.1 M): Na_2HPO_4 (15.46 g/L), NaH_2PO_4 (5.83 g/L).

2.4 Equipment and Software

1. Compact refrigerated centrifuge (Eppendorf, USA).
2. Dialysis cassette (Slide-a-Lyzer, 3,000 MWCO, 0.5–3 ml capacity, Thermo Scientific, USA).
3. Dry bath heater (Daigger, USA).
4. Ultrafree centrifugal filter units (EMD Millipore, USA).
5. Flow cytometry (FACScan Cytofluorimeter, Becton Dickinson USA) and FlowJo software (TreeStar, USA).
6. ImageJ software (NIH, USA).

3 Methods

3.1 Bone Marrow-Derived MC Differentiation

1. Obtain the BMMCs by in vitro differentiation of bone marrow precursors from mouse femurs by repetitive bone marrow flushing with BMMC culture medium. Grow the precursors in BMMC culture medium containing 20 ng/ml of IL-3 in a humidified 37 °C, 5 % CO_2 incubator for 4–5 weeks. Change the medium twice a week and adjust the cell density to 1×10^6 cells/ml.
2. At week 4, evaluate the purity of BMMC by measuring FcεRI expression by flow cytometry (anti-mouse FcεRI antibody). After 5 weeks, BMMCs are usually more than 98 % FcεRI-positive stained.

3.2 sOX40 Purification

OX40-IgG1 fusion protein contains mouse OX40 extracellular domain (residues 23–198) in fusion with a mouse IgG1 Fc. Remove the Fc portion of OX40-IgG1 (hinge, CH2, CH3) by digestion with papain. Treat the solution containing Fc and OX40 (sOX40) fragments with iodoacetamide in order to inactivate the enzymatic activity of papain. Apply the solution containing Fc and OX40 to affinity chromatography and the Fc fragment is retained in the column (*see* Fig. 1). Steps for sOX40 purification are detailed here.

1. Adjust the concentration of the OX40-IgG fusion protein to 2 mg/ml in digestion buffer.
2. Add 10 μg of papain/mg of OX40-IgG1.
3. Allow digestion occur for 16–18 h at 37 °C using a dry bath heater (*see* **Note 1**).
4. Incubate the mix with 0.3 M iodoacetamide (final volume) for 1 h at room temperature.
5. Wash the Protein G Sepharose two times with PBS, centrifuge at $800 \times g$ for 5 min at 4 °C, discard the supernatant and load the protein mixture on the Protein G Sepharose (ratio 3:1 v/v mix:resin volume), and incubate for 2 h at 4 °C to separate the Fc portion.

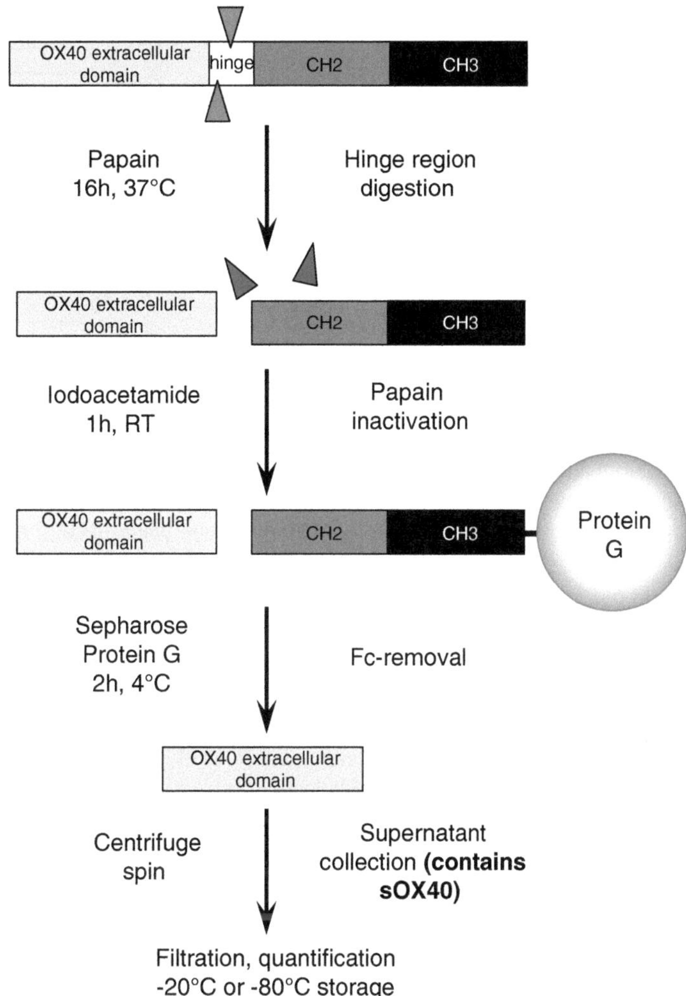

Fig. 1 Purification of the sOX40 molecule. Schematic representation of digestion, papain inactivation, and Fc removal from the murine OX40-Fc fusion protein

6. After centrifugation, collect the supernatant (sOX40 is in the supernatant) and discard the resin.
7. Load the supernatant onto a dialysis cassette and dialyze the eluted fraction overnight in 0.1 M phosphate buffer. This step will dialyze the iodoacetamide present in the solution.
8. Collect and filtrate the supernatant in Ultrafree centrifugal filter units (*see* **Note 2**).

3.3 Quantification

Load the eluted fraction onto a polyacrylamide gel and analyze with Coomassie Brilliant Blue staining. Perform the quantification through densitometric analysis of the stained band and compare to one standard with known concentration. ImageJ software is used for densitometric quantification.

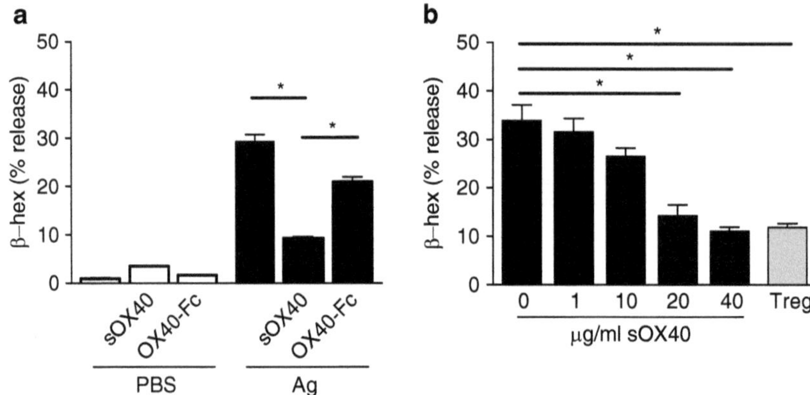

Fig. 2 Mast cell degranulation assay for sOX40 molecule. (**a**) WT IgE-presensitized BMMCs were challenged with 50 ng/ml antigen (Ag) in the absence or presence of 20 μg/ml sOX40 or undigested molecule (OX40-Fc) and were examined for β-hexosaminidase (β-hex) release. The purification of the sOX40 increases the effectiveness of the molecule to reduce BMMC degranulation when BMMCs are triggered with Ag and prevents nonspecific interactions (−70 % BMMC degranulation compared to Ag-triggered BMMCs in the absence of sOX40, (**a**) *black bars*). No significant differences in β-hex release are found when BMMCs are treated in the absence of Ag stimulation (*white columns*) (**b**) sOX40 concentration and inhibition of BMMC degranulation. BMMCs were stimulated with 50 ng/ml Ag in the presence of increasing concentrations of sOX40 (1, 10, 20, and 40 μg/ml, *black columns*). As positive control of the experiment, BMMCs were challenged with 50 ng/ml Ag in the presence of the same number of OX40-expressing CD4+/CD25+ regulatory T cells (*grey column*) in the same final volume. Graphs show means ± S.E.M. of at least two independent experiments, each of which gave similar results. $^*P<0.01$, using a Student's *t* test

3.4 BMMC Degranulation Assay for the Functional Characterization of sOX40

We routinely tested the efficacy of sOX40 purification with a BMMC degranulation test (*see* Fig. 2, **Notes 3** and **4**).

1. Before all in vitro experiments, starve the BMMCs 1 h in low serum, cytokine-free culture medium. Then sensitize 1×10^6/ml BMMCs at 37 °C for 3 h with 1 μg of anti-DNP specific IgE.
2. Wash the BMMCs, resuspend in 500 μl of Tyrode's buffer and challenge with 50 ng/ml of DNP in the absence or presence of 20 μg/ml of sOX40 at 37 °C for 30 min.
3. To stop degranulation, incubate the BMMCs at 4 °C for 10 min. Then, centrifuge at 4 °C, 200 × *g* for 5 min and collect the supernatant, which contains the amount of released β-hexosaminidase (β-hex, 500 μL) (supernatant 1).
4. Lyse the BMMC pellet in the same volume (500 μL) of lysis buffer at 4 °C for 15 min and centrifuge at 20,000 × *g* for 1 min. Collect the resulting supernatant, which contains the amount of β-hex retained in the cell cytoplasm (supernatant 2).
5. Using a flat bottom 96-well plate, add 50 μL of p-NAG buffer to the same volume of supernatants 1 and 2 and incubate the

plate at 37 °C for 1 h. The β-hex present in the supernatants cleaves p-NAG into p-nitrophenol.

6. Then stop the reaction by adding 200 μL of carbonate buffer to each well. The pH shift of the reaction determines the formation of a chromogenic product, which intensity is proportional to the amount of β-hex contained in supernatants.
7. Measure the amount of chromogenic substrate using an ELISA reader at 405 nm wavelength. Record the optical density (O.D.) and calculate the percentage of release with the following formula:

$$\frac{\text{O.D.supernatant 1}}{(\text{O.D.supernatant 1} + \text{O.D.supernatant 2})} \times 100$$

4 Notes

1. Overdigestion of the sample should be avoided, since papain can cleave cysteines present in the OX40 portion, aspecifically. If overdigestion occurs, load the sample onto a polyacrylamide gel (8 %), stain the gel with Coomassie Brilliant Blue Staining, and check for aspecific bands (<20 kDa). Repeat the entire purification process with a new OX40-IgG batch, if aspecific bands are detected.
2. Filtered sOX40 aliquots can be stored at −80 °C up to 1 year. Avoid repeated freeze/thaw cycles.
3. The degranulation test measures the extent of MC response to the Ag, evaluating the amount of the granule-associated enzyme β-hex released from IgE-sensitized, Ag-triggered BMMCs.
4. We used the sOX40 to study the impact of OX40L engagement to several FcεRI-dependent events in MCs, such as phosphorylation/dephosphorylation of signaling proteins, calcium influx, microtubule organization, cytokine and chemokine production.

   The sOX40 is a useful tool to dissect and characterize the contribution of OX40L-mediated signaling to MC activation (and potentially to all OX40L-expressing cell types activation), with a simpler experimental design, which does not require the use of co-culture systems. Thus, the sOX40 represents a powerful tool, virtually applicable to the study of OX40L-expressing populations, such as activated B cells, mature conventional dendritic cells (DCs), Langerhans cells, plasmacytoid DCs, and macrophages [15].

References

1. Galli SJ, Tsai M (2012) IgE and mast cells in allergic disease. Nat Med 18:693–704
2. Gilfillan AM, Austin SJ, Metcalfe DD (2011) Mast cell biology: introduction and overview. Adv Exp Med Biol 716:2–12
3. Gri G, Frossi B, D'Inca F, Danelli L, Betto E, Mion F et al (2012) Mast cell: an emerging partner in immune interaction. Front Immunol 3:120
4. Nakae S, Suto H, Iikura M, Kakurai M, Sedgwick JD, Tsai M et al (2006) Mast cells enhance T cell activation: importance of mast cell costimulatory molecules and secreted TNF. J Immunol 176:2238–2248
5. Frossi B, D'Inca F, Crivellato E, Sibilano R, Gri G, Mongillo M et al (2011) Single-cell dynamics of mast cell-CD4+ CD25+ regulatory T cell interactions. Eur J Immunol 41: 1872–1882
6. Miura S, Ohtani K, Numata N, Niki M, Ohbo K, Ina Y et al (1991) Molecular cloning and characterization of a novel glycoprotein, gp34, that is specifically induced by the human T-cell leukemia virus type I transactivator p40tax. Mol Cell Biol 11:1313–1325
7. Baum PR, Gayle RB, Ramsdell F, Srinivasan S, Sorensen RA, Watson ML et al (1994) Molecular characterization of murine and human OX40/OX40 ligand systems: identification of a human OX40 ligand as the HTLV-1-regulated protein gp34. EMBO J 13:3992–4001
8. Kashiwakura J, Yokoi H, Saito H, Okayama Y (2004) T cell proliferation by direct cross-talk between OX40 ligand on human mast cells and OX40 on human T cells: comparison of gene expression profiles between human tonsillar and lung-cultured mast cells. J Immunol 173: 5247–5257
9. Takeda I, Ine S, Killeen N, Ndhlovu LC, Murata K, Satomi S et al (2004) Distinct roles for the OX40–OX40 ligand interaction in regulatory and nonregulatory T cells. J Immunol 172:3580–3589
10. Gri G, Piconese S, Frossi B, Manfroi V, Merluzzi S, Tripodo C et al (2008) CD4+ CD25+ regulatory T cells suppress mast cell degranulation and allergic responses through OX40-OX40L interaction. Immunity 29:771–781
11. Piconese S, Gri G, Tripodo C, Musio S, Gorzanelli A, Frossi B et al (2009) Mast cells counteract regulatory T-cell suppression through interleukin-6 and OX40/OX40L axis toward Th17-cell differentiation. Blood 114: 2639–2648
12. Sibilano R, Frossi B, Suzuki R, D'Incà F, Gri G, Piconese S et al (2012) Modulation of FcεRI-dependent mast cell response by OX40L via Fyn, PI3K, and RhoA. J Allergy Clin Immunol 130:751–760
13. Sibilano R, Gri G, Frossi B, Tripodo C, Suzuki R, Rivera J et al (2011) Technical advance: soluble OX40 molecule mimics regulatory T cell modulatory activity on FcεRI-dependent mast cell degranulation. J Leukoc Biol 90: 831–838
14. Tkaczyk C, Okayama Y, Metcalfe DD, Gilfillan AM (2004) Fcgamma receptors on mast cells: activatory and inhibitory regulation of mediator release. Int Arch Allergy Immunol 133:305–315
15. Croft M (2010) Control of immunity by the TNFR-related molecule OX40 (CD134). Annu Rev Immunol 28:57–78

Chapter 4

Analyzing the Signaling Capabilities of Soluble and Membrane TWEAK

Johannes Trebing, José Antonio Carmona Arana, Steffen Salzmann, and Harald Wajant

Abstract

TWEAK, like many other ligands of the TNF family, occurs naturally in two forms, as a type II transmembrane protein and as soluble ligand released from the latter by proteases of the furin family. Both TWEAK variants interact with high affinity with Fn14, an unusual small member of the TNF receptor family. TWEAK and Fn14 activate a variety of intracellular signaling pathways but regulation of TNF-induced cell death and stimulation of the classical and alternative NFκB pathway are certainly the best understood ones. Intriguingly, soluble and membrane TWEAK significantly differ in their ability to trigger these responses. While activation of the alternative NFκB pathway and enhancement of TNF-induced cell death are efficiently induced by both forms of TWEAK, membrane TWEAK has a much higher capacity than soluble TWEAK to stimulate the classical NFκB pathway. Importantly, soluble TWEAK gains a membrane TWEAK-like Fn14 stimulating activity upon oligomerization or artificial anchoring to the cell surface. On the example of NFκB signaling and enhancement of TNF-induced cell death, we summarize here protocols that allow the identification of signaling pathways/cellular responses that preferentially respond to membrane TWEAK. These protocols base either on the side-by-side analysis of soluble TWEAK and oligomerized or cell surface-anchorable TWEAK variants or on the use of transfectants expressing soluble and membrane TWEAK.

Key words Fn14, Classical and alternative NFκB pathway, Soluble and membrane TWEAK, TNF

1 Introduction

TWEAK as the huge majority of the other TNF ligand family members naturally occurs in two forms, as a type II transmembrane protein and as a soluble molecule that is released from the latter by proteolytic processing [1, 2]. Both soluble and membrane-bound TWEAK contain the characteristic C-terminal TNF homology domain and thus assemble into homotrimeric proteins with the capability to interact with Fn14, a member of the TNF receptor superfamily. Importantly, while all TNF receptors (with exception of the decoy TNF receptors) become comprehensively activated by interaction with their membrane-bound TNF ligands,

Jagadeesh Bayry (ed.), *The TNF Superfamily: Methods and Protocols*, Methods in Molecular Biology, vol. 1155, DOI 10.1007/978-1-4939-0669-7_4, © Springer Science+Business Media New York 2014

they differ in their response to stimulation with soluble ligand trimers. Some members of the TNF receptor family, e.g., TNFR1, are strongly responsive for their soluble ligand whereas others, e.g., TNFR2, CD95, TRAILR2, OX40, 41BB, CD27, and TACI admittedly bind their receptor but without inducing a robust response [3–8]. In this respect Fn14 is quite unusual as the stimulating effect of binding of soluble TWEAK is dependent on the signaling pathway considered. Fn14-mediated activation of the alternative NFκB pathway and enhancement of TNF-induced cell death are efficiently induced by soluble and membrane-bound TWEAK as well as by artificial soluble TWEAK variants mimicking the activity of membrane TWEAK [9–11]. In contrast to the other TWEAK forms, however, soluble TWEAK trimers largely fail to trigger TRAF2 degradation and activation of the classical NFκB pathway [9–11]. The limited responsiveness of Fn14 to soluble TWEAK can be overcome in two ways, first, by oligomerization of soluble TWEAK trimers [10, 11]. So, Flag-tagged soluble TWEAK crosslinked with anti-Flag and dimers of TWEAK trimers generated by genetic engineering induce activation of the classical NFκB pathway with more than 100-fold lower ED50-values than soluble TWEAK trimers, but all these TWEAK variants trigger the alternative NFκB pathway with comparable efficacy [11]. Noteworthy, oligomerization of TWEAK trimers has no major effect on the dose dependency of Fn14 occupancy indicating that the enhancing effect of ligand oligomerization in context of Fn14-mediated classical NFκB signaling is not related to an avidity-driven increase in apparent affinity. Instead this points to a need of secondary interaction of initially formed trimeric ligand receptor complexes for full Fn14 activation [12]. The second mean to achieve a comprehensive Fn14 response with soluble TWEAK trimers is their artificial immobilization on the cell surface. This can be achieved by fusing soluble TWEAK by genetic engineering N-terminally with a single-chain variable fragment (scFv) antibody fragment with specificity for a cell surface-exposed antigen. If there is no antigen present for binding, this type of scFv-TWEAK fusion proteins behaves similar to conventional soluble TWEAK trimers and thus has no major stimulatory effect on the classical NFκB pathway. If such scFv-TWEAK fusion proteins, however, have the opportunity to anchor to the cell surface of antigen expressing cells via their scFv domain, they exhibit membrane TWEAK-like activities and, for example, strongly activate the classical NFκB pathway [11]. Thus, cellular responses that have already been efficiently triggered by soluble TWEAK can be recognized by the fact that they are induced with similar dose dependency and amplitude by soluble TWEAK and scFv-TWEAK irrespective of cell surface-antigen binding. In contrast, the induction of Fn14 responses requiring membrane TWEAK by scFv-TWEAK fusion proteins is much stronger and more efficient than those triggered by soluble TWEAK when such variants have the opportunity to anchor to their cell surface antigen.

In this chapter, we describe on the one hand a step-by-step protocol for the production of soluble Flag-TWEAK, Fc-Flag-TWEAK, and a scFv-Flag-TWEAK fusion protein with specificity for the cell surface antigen FAP. By taking the examples of the classical and alternative NFκB pathway and enhancement of TNF-induced cell death, we provide on the other hand procedures to use the recombinant TWEAK proteins, or cell transfectants expressing soluble and membrane TWEAK, to figure out whether a certain Fn14 response is differently triggered by the soluble and membrane-bound form of TWEAK.

2 Materials

2.1 Production and Purification of Flag-TWEAK, Fc-Flag-TWEAK, and scFv-Flag-TWEAK Fusion Proteins

1. Eukaryotic expression plasmid (e.g., pCR3 based) encoding the TWEAK variant of interest.
2. HEK293 cells (ATCC, USA).
3. Complete RPMI 1640 medium: RPMI 1640, 10 % fetal calf serum (FCS) (PAA, Germany).
4. Easyject Plus Electroporator and 4 mm cuvettes (PeqLab, Germany).
5. 15 cm tissue culture dishes (Greiner Bio-One, Germany).
6. Anti-Flag mAb M2 agarose (Sigma, Germany).
7. Tris buffer.
8. TBS: 0.02 M Tris–HCl, pH 7.6, 8 % NaCl.
9. Glycerol 50 % in TBS with 0.02 % sodium azide.
10. Elution buffer: TBS, 100 μg/mL Flag peptide (Sigma, Germany).

2.2 Biochemical Characterization

1. Standard buffers and equipment for SDS-PAGE, Western blotting, ELISA, and Gel filtration chromatography.
2. Broad range Prestained Protein Marker Kit (New England Biolabs, Germany).
3. Low Molecular Weight Calibration Kit for SDS Electrophoresis ("silver gel marker," GE Healthcare, Germany).
4. Nitrocellulose membrane (Schleicher und Schuell, Germany).
5. Page Silver, Silver Staining Kit (Fermentas GmbH, Germany).
6. Anti-Flag M2, mouse IgG1 mAb (Sigma, Germany) or anti-TWEAK (R&D Systems, Germany).
7. Polyclonal rabbit anti-mouse Immunoglobulins/HRP (Dako-Cytomation, Denmark).
8. ECL detection system (Amersham Biosciences, Germany).
9. TWEAK ELISA (R&D Systems, Germany).
10. Phosphate buffered saline (PBS): 137 mM NaCl, 2.7 mM KCl, 8 mM Na_2HPO_4, 2 mM KH_2PO_4.

11. Gel filtration column BioSep-SEC-S3000 (300×7.8) (Phenomenex, Germany).
12. Column performance check standard aqueous SEC 1 solution (Phenomenex, Germany).

2.3 Analysis of the Relevance of Soluble Versus Membrane TWEAK for Fn14-Mediated NFκB Activation

1. NCTC 2472 (ATCC, USA).
2. Fn14-expressing human cell lines, e.g., HT1080, Colo 205, HT29, or SKOV3 (ATCC, USA).
3. Standard buffers and equipment for SDS-PAGE, Western blotting, ELISA, and FACS analysis.
4. Broad range Prestained Protein Marker Kit.
5. Nitrocellulose membrane.
6. Anti-NIK (Cell Signaling, Germany).
7. Anti-p100/p52 (Upstate Biotech, Germany).
8. Anti-TRAF2 (Santa Cruz Biotechnologies or BD Biosciences, Germany).
9. Anti-cIAP1 and anti-cIAP2 (Cell Signaling).
10. Anti-IκBα (Santa Cruz Biotechnologies) and anti-phospho IκBα (Cell Signaling).
11. HRP-conjugated secondary Abs (Dako).
13. ECL detection system.
12. Lipofectamine 2000 (Invitrogen, Germany).
13. TWEAK ELISA.
14. Anti-TWEAK and control goat IgG (R&D Systems).
15. PE-labeled goat IgG-specific porcine Ab (R&D Systems).
16. IL-8 or IL-6 ELISA (R&D Systems).
17. PBS.
18. Crystal violet (Sigma, Germany).
19. 3-[4,5-dimethylthiazol-2-yl]-2,5-diphenyl tetrazolium bromide (MTT) (Sigma, Germany).

3 Methods

The methods used for production and purification of the various TWEAK variants are basically similar to those recently published by us for scFv-TNF ligand fusion proteins [13]. The evaluation of the question whether a particular type of Fn14-mediated response is differentially stimulated by soluble and membrane TWEAK is exemplified here on the example of the classical and alternative NFκB pathway and enhancement of TNF-induced cell death. Essentially the same experimental setups can be used to evaluate the relevance of soluble and membrane TWEAK for the engagement of other Fn14 responses (*see* **Note 1**) when the analysis of an appropriate read out is established.

3.1 Production and Purification of Flag-TWEAK, Fc-Flag-TWEAK, and scFv-Flag-TWEAK Fusion Proteins

3.1.1 Construction of Eukaryotic Expression Plasmids

1. Construct by help of standard cloning techniques eukaryotic expression plasmids encoding the TWEAK variants of interest with a Flag tag.
2. Use the following domain architectures: For Flag-TWEAK: Leader sequence—Flag tag—TWEAK domain, for Fc-Flag-TWEAK: Leader sequence—hIgG1 Fc domain—Flag tag—TWEAKdomainandforscFv-Flag-TWEAK:Leader—scFv—Flag tag—TWEAK domain (*see* **Notes 2–4**).
3. The TWEAK domain must contain the TNF homology domain to allow receptor binding and trimerization.

3.1.2 Transient Expression in HEK293 Cells

1. Harvest a confluent 175 cm^2 flask of HEK293 cells (app. $40–50 \times 10^6$ cells) and wash cells once with RPMI 1640 medium.
2. Resuspend cells in 1 mL of complete RPMI 1640 medium.
3. Add 30–40 μg of the expression construct of interest (in TE or H_2O) in a maximal volume of 50 μL.
4. Electroporate (250 V, 1,800 μF, maximum resistance) the cell/plasmid mixture in a 4 mm cuvette using an Easyject Plus electroporator.
5. Recover the electroporated cells in 20 mL of complete RPMI 1640 medium in a 15 cm tissue culture dish overnight.
6. Aspirate medium to remove dead cells and add fresh RPMI 1640 medium with low FCS (0.5–2 % FCS).
7. Collect cell culture supernatant after 4–6 days and determine the concentration of the TWEAK variant, e.g., by anti-Flag Western blotting and a Flag-tagged protein standard of known concentration or using a TWEAK ELISA. Be aware that a significant fraction of the cells detach under the low serum cultivation conditions. The yields obtained with this method are typically between 10 and 15 μg/mL for Flag-TWEAK and 3–6 μg/mL for Flag-Fc-TWEAK. The concentrations of scFv-Flag-TWEAK fusion proteins that can be reached are highly dependent on the scFv domain used and range from <100 ng/mL up to 5 μg/mL (*see* **Note 5**).

3.1.3 Affinity Purification on Anti-Flag Agarose

1. Filtrate the TWEAK variant-containing supernatant through a 0.2 μm filter.
2. Suspend anti-Flag mAb M2 agarose (1 mL per 500 μg of Flag-tagged TWEAK variant) and transfer the suspension to a clean empty column.

3. Allow the anti-Flag mAb M2 agarose to settle (minimal bed volume 200 μL, minimal bed high 3× diameter).
4. Drain the anti-Flag mAb M2 agarose column to the top of the agarose bed but letting it not go dry.
5. Equilibrate the column twice with 5–10 agarose bed volumes of TBS. Again prevent the agarose bed from running dry.
6. Supplement the TWEAK-containing cell culture supernatant with NaCl to reach a final concentration of 150 mM and load it onto the column using gravity flow and a flow rate of 0.5–1 mL/min. Do not disturb the agarose bed while loading.
7. Remove unbound proteins by washing the agarose with three times 5 column volumes of TBS.
8. Drain the anti-Flag mAb M2 agarose column again to the top of the gel bed without letting it go dry.
9. Elute the Flag-tagged TWEAK variant with 6× one column volume of Flag peptide elution buffer. Drain the column completely after application of each aliquot and load the elution buffer without disturbing the agarose bed.
10. Prepare anti-Flag mAb M2 agarose for storage and reuse by washing it with 10 column volumes of TBS-glycerol buffer. Store anti-Flag mAb M2 agarose at −20 °C (*see* **Note 6**).
11. Determine concentration of the purified TWEAK variant using a commercially available protein concentration determination kit or a TWEAK ELISA. The TWEAK concentration may also be determined in course of the biochemical characterization of the protein by SDS-PAGE and silver staining and comparison with corresponding standard proteins.
12. Sterile filtrate the protein solution.

3.2 Biochemical Characterization

3.2.1 SDS-PAGE and Silver Staining

1. To evaluate the purity of the affinity chromatography purified TWEAK proteins, resolve samples and molecular weight standards under reducing conditions on a 12 % SDS-PAGE gel.
2. Stain the resolved proteins using a commercially available silver staining kit.
3. SDS-PAGE analysis and silver staining of the anti-Flag affinity purified TWEAK variants might also be used to determine their concentration by comparison with the bands of the protein standard used.

3.2.2 Gel Filtration Analysis

1. Equilibrate gel filtration column with PBS and determine the elution volume of a set of appropriate standard proteins.
2. Analyze the purified TWEAK variant (50–500 μg/mL) using flow rate and sample volume as specified for the gel filtration column used.

3. If there are aggregated or high molecular weight TWEAK species, check the activity of the corresponding fraction(s) for IL-8 induction (*see* Subheading 3.3.1).
4. If there is significant activity, the aggregates might be removed by preparative gel filtration to improve the quality of the preparation regarding its oligomerization- or cell surface-anchorable-dependent activity (*see* **Note 7**).

3.3 Analysis of the Different Signaling Capabilities of Soluble and Membrane TWEAK Using Flag-TWEAK and Anti-Flag Oligomerization

3.3.1 Analysis of IL-8 or IL-6 Induction

1. Seed Fn14-expressing cells (e.g., HT1080, Colo 205, or HT29 cells, 10–20 × 10^3 cells per well in 100 μL) in 96-well tissue culture plates.
2. Next day, prepare on a separate 96-well plate 6 rows of serial dilutions of Flag-TWEAK in cell culture medium. Start, e.g., with a concentration of 1,000 ng/mL and dilute 1–5. Do not add Flag-TWEAK to the last column (negative control).
3. Remove the medium from the cells prepared the day before and transfer the Flag-TWEAK solutions (100 μL).
4. On one half of the rows add 100 μL of anti-Flag mAb M2 (1 μg/mL) to oligomerize Flag-TWEAK, on the other half add 100 μL of medium. Final concentration of M2 in the assay is 0.5 μg/mL and the highest Flag-TWEAK concentration is 500 μg/mL.
5. Incubate overnight (12–18 h) in a CO_2 incubator at 37 °C.
6. Analyze supernatants with respect to their IL-8 or IL-6 content. Both are dominantly controlled by the classical NFκB pathway and their induction is typically tightly correlated with the activity of this pathway.
7. Plot the dose response of IL-8/IL-6 induction of Flag-TWEAK and oligomerized Flag-TWEAK. Oligomerization of Flag-TWEAK lowers the concentrations required to trigger IL-8/IL-6 production typically for more than two orders of magnitude and indicate the superior classical NFκB-stimulating activity of membrane TWEAK (Fig. 1).

3.3.2 Analysis of Enhancement of TNF-Induced Cell Death

1. Seed Fn14-expressing cells (e.g., HT1080, Colo 205, or HT29 cells, 10–20 × 10^3 cells per well in 100 μL) in 96-well tissue culture plates (*see* **Note 8**).
2. Next day, prepare on a separate 96-well plate 6 rows of serial dilutions of Flag-TWEAK in cell culture medium. Start with a concentration of 2,000 ng/mL and dilute 1–5. Do not add Flag-TWEAK to the last row (negative control).
3. Replace the medium on the cells prepared the day before with medium containing 100 ng/mL of TNF and 5 μg/mL of CHX. As most cells are fairly resistant against TNF/TWEAK, this treatment is required to sensitize cells and ensure robust cell death induction. Please be aware that despite CHX treatment most cells nevertheless poorly respond to TNF alone.

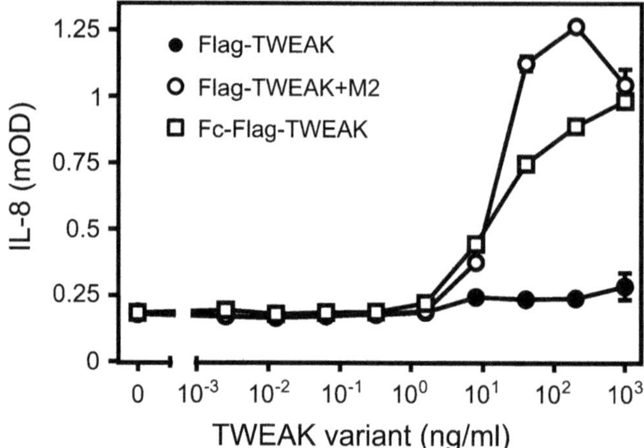

Fig. 1 Superior IL8 induction by oligomerized variants of soluble TWEAK. Colo 205 cells were challenged in triplicates as described under Subheading 3.3.1 with the indicated concentrations of Flag-TWEAK, oligomerized Flag-TWEAK, and Fc-Flag-TWEAK. Next day, supernatants were analyzed with respect to IL-8 production

4. Transfer 50 μL of the diluted Flag-TWEAK solutions to the TNF/CHX challenged cells.
5. On one half of the rows add 50 μL of anti-Flag mAb M2 (2 μg/mL) to oligomerize Flag-TWEAK, on the other half add 50 μL of medium. Final concentration of TNF in the assay is 50 ng/mL, the highest Flag-TWEAK concentration is 500 ng/mL, and anti-Flag M2 and CHX reach concentrations of 0.5 μg/mL and 2.5 μg/mL.
6. Incubate for 24 h in a CO_2 incubator at 37 °C.
7. Determine cellular viability by crystal violet staining or using MTT.
8. Plot the cellular viability as a function of the concentration of Flag-TWEAK and oligomerized Flag-TWEAK. Oligomerization has typically no major effect on the dose response relation of TWEAK-mediated enhancement of TNF-induced cell death.

3.3.3 Analysis of NFκB Signaling by Western Blotting

1. Seed Fn14-expressing cells (e.g., HT1080, WiDr, or HT29 cells, 10–20 × 10^3 cells per well in 100 μL) in 6-well tissue culture plates.
2. Next day, stimulate cells with an appropriate series of concentrations (e.g., 500, 100, 20, 4, 0.8, 0.016 ng/mL) of Flag-TWEAK and oligomerized Flag-TWEAK. The latter is obtained by preincubation of corresponding tenfold Flag-TWEAK solutions with 5 μg/mL anti-Flag mAb M2.

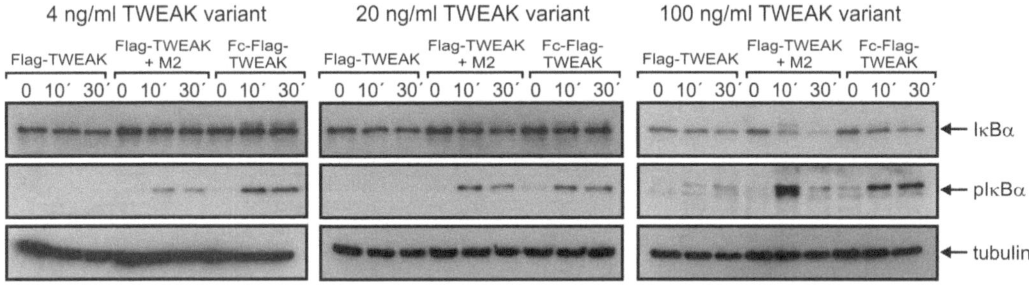

Fig. 2 Oligomerization strongly enhances the capability of soluble TWEAK to trigger phosphorylation and degradation of IκBα. HT1080 cells were stimulated for 10 and 30 min with 4, 20, and 100 ng/mL of Flag-TWEAK, oligomerized Flag-TWEAK, and Fc-Flag-TWEAK or remained untreated. Total cell lysates were prepared and analyzed by Western blotting for the presence of IκBα and pospho-IκBα

3. For analysis of alternative NFκB signaling challenge cells for 6–8 h. For analysis of classical NFκB signaling stimulate cells for 10 and 30 min.
4. Harvest cells using a rubber policeman and wash cells twice with PBS.
5. Collect cells by centrifugation, resuspend them in 4× Laemmli sample buffer, and lyse them by sonification (10 pulses) and boiling (5 min, 96 °C).
6. Subject samples dedicated for analysis of alternative NFκB signaling to SDS-PAGE and Western blot analysis of expression of NIK and p100/p52 (*see* **Note 9**). Samples dedicated for analysis of classical NFκB signaling will be analyzed similarly with respect to the presence of IκBα and pospho-IκBα.
7. Differences in the dose response relation between Flag-TWEAK and oligomerized Flag-TWEAK are again indicative for effects preferentially triggered by membrane TWEAK (Fig. 2).

3.4 Analysis of the Different Signaling Capabilities of Soluble and Membrane TWEAK Using Flag-TWEAK and Fc-Flag-TWEAK

3.4.1 Analysis of IL-8 or IL-6 Induction

1. Seed Fn14-expressing cells (e.g., HT1080, WiDr, or HT29 cells, 10–20 × 10^3 cells per well in 100 μL) in 96-well tissue culture plates.
2. Next day, prepare on a separate 96-well plate in 3 rows of serial dilutions of Flag-TWEAK and in 3 other rows corresponding dilutions of Fc-Flag-TWEAK. Start with a concentration of 500 ng/mL and dilute 1–5. Use the last row as negative control (no TWEAK).
3. Remove the medium from the cells prepared the day before and transfer the TWEAK solutions (100 μL).
4. Proceed as described under **steps 5–7** in Subheading 3.3.1 (Fig. 1).

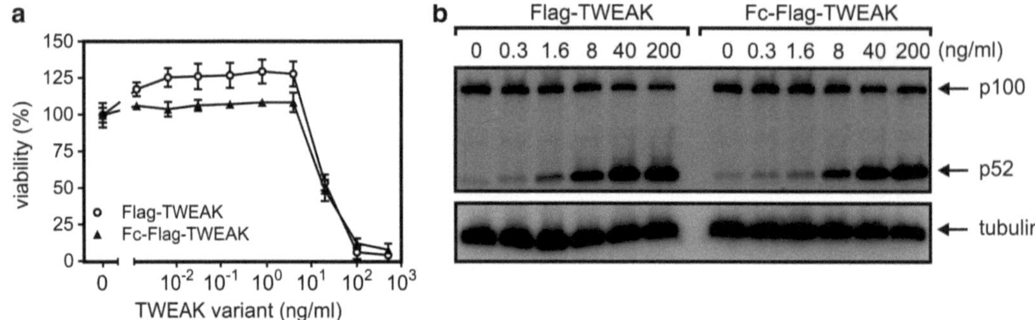

Fig. 3 Oligomerization has no relevant enhancing effect on induction of cell death and alternative NFκB signaling by soluble TWEAK. (**a**) SKOV3 cells, which undergo endogenous-TNF-mediated Fn14-induced cell death, were challenged in triplicates with the indicated concentrations of Flag-TWEAK and Fc-Flag-TWEAK. Next day, cellular viability was determined by crystal violet staining and normalized against cells that were killed using a highly toxic mixture containing CHX, sodium azide, and Fc-CD95L. (**b**) HT29 cells were stimulated with increasing concentrations of Flag-TWEAK and Fc-Flag-TWEAK. Total cell lysates were analyzed by Western blotting with respect to p100 processing

3.4.2 Analysis of Enhancement of TNF-Induced Cell Death

1. Seed Fn14-expressing cells (e.g., HT1080, WiDr, or HT29 cells, $10–20 \times 10^3$ cells per well in 100 μL) in 96-well tissue culture plates (*see* **Note 8**).
2. Next day, prepare on a separate 96-well plate in 3 rows of serial dilutions of Flag-TWEAK and in 3 other rows corresponding dilutions of Fc-Flag-TWEAK. Start with a concentration of 1,000 ng/mL and dilute 1–5. Use the last row as negative control (no TWEAK).
3. Replace the medium on the cells prepared the day before with medium containing 100 ng/mL of TNF and 5 μg/mL of CHX.
4. Transfer 100 μL of the diluted Flag-TWEAK and Fc-Flag-TWEAK samples to the TNC/CHX challenged cells.
5. Proceed as described under **steps 6–8** in Subheading 3.3.2 (Fig. 3a).

3.4.3 Analysis of NFκB Signaling by Western Blotting

1. Seed Fn14-expressing cells (e.g., HT1080, WiDr, or HT29 cells, $10–20 \times 10^3$ cells per well in 100 μL) in 6-well tissue culture plates.
2. Next day, stimulate cells with an appropriate series of concentrations (e.g., 500, 100, 20, 4, 0.8, 0.016 ng/mL) of Flag-TWEAK and Fc-Flag-Fc-TWEAK.
3. Proceed as described under **steps 3–7** in Subheading 3.3.3 (Fig. 3b).

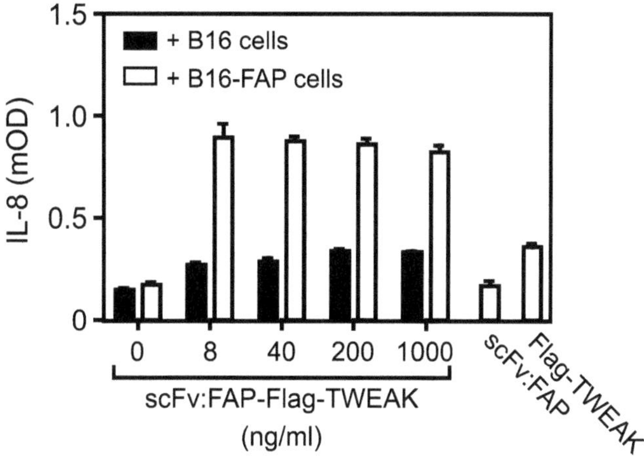

Fig. 4 Cell surface anchoring of soluble TWEAK potentiates its IL8-inducing activity. Cocultures of B16 and B16-FAP cells with HT29 cells were stimulated in triplicates with the indicated concentrations of scFv:FAP-Flag-TWEAK as described under Subheading 3.5.1 and supernatants were analyzed the next day for their IL-8 content by ELISA. As a control supernatants of B16-FAP/HT29 cocultures treated with 1,000 ng/mL Flag-TWEAK or 1,000 ng/mL scFv:FAP were included

3.5 Analysis of the Relevance of Soluble Versus Membrane TWEAK for Fn14-Mediated NFκB Activation Using a TWEAK Fusion Protein That Anchor to the Cell Surface Expressed FAP Molecule

3.5.1 Analysis of IL-8 or IL-6 Induction

1. Seed FAP-negative murine B16 cells (3 rows) and corresponding FAP transfectants (3 rows) in 96-well tissue culture plates in coculture with Fn14-expressing human cells.
2. Next day, prepare on a separate 96-well plate 6 rows of serial dilutions of scFv:FAP-Flag-TWEAK, starting with a concentration of 100–500 ng/mL and using 1–5 dilution steps. Use the last row as negative control (no scFv:FAP-Flag-TWEAK).
3. Remove the cell culture medium from the cells prepared the day before to minimize the background of constitutive IL-8 production and incubate the cells with 100 μL per well of the serial dilutions of scFv:FAP-Flag-TWEAK (*see* **Note 10**).
4. Proceed as described under **steps 5–7** in Subheading 3.3.1 (Fig. 4).

3.5.2 Analysis of NFκB Signaling by Western Blotting

1. Seed cocultures of FAP-negative murine B16 cells and corresponding FAP transfectants with human Fn14-expressing cells (e.g., HT1080, WiDr, or HT29 cells, $10–20 \times 10^3$ cells per well in 100 μL) in a ratio of 1:2 to 1:3 in 6-well tissue culture plates.
2. Next day, stimulate cocultures with various concentrations (e.g., 500, 100, 20, 4, 0.8, 0.016 ng/nl) of scFv:FAP-Flag-TWEAK.
3. Proceed as described under **steps 3–7** in Subheading 3.3.3.

3.6 Analysis of Fn14-Mediated NFκB Activation by Soluble and Membrane TWEAK Expressing Cells

1. Transfect murine NCTC 2472 cells (*see* **Note 11**) two consecutive days with expression constructs encoding Flag-TWEAK and full-length TWEAK or empty vector using Lipofectamine 2000 according to the protocol recommended by the supplier.
2. On day 3, harvest the transfected NCTC 2472 cells and analyze for cell surface expression of TWEAK by FACS. Analyze the supernatants in parallel with respect to their soluble TWEAK content. To proceed with functional analysis, the membrane TWEAK expressing cells should be >50 % positive (*see* **Note 12**).
3. Mix the various NCTC 2472 transfectants 1:1 with human Fn14-expressing cell lines and seed cells in triplicates in a 96-well plate or in a 6-well plate. Ensure that cell density is high enough to guarantee effective cell-to-cell interactions.
4. On day 4, analyze the supernatants of the 96-well cocultures by IL-8 or IL-6 ELISA
5. Harvest the 6-well cocultures and analyze by Western blotting for activity of the alternative NFκB pathway (NIK accumulation, p100 processing).
6. To analyze the effect of the various transfectants on the classical NFκB pathway (IκBα, phospho-IκBα), they are directly used after harvesting to stimulate Fn14-expressing cells seeded the day before in a 6-well plate (time course of 0–2 h). Also test the supernatants in parallel (*see* **Note 13**).

4 Notes

1. An interaction of TWEAK with the scavenger receptor CD163 has been published [13] but we failed to confirm this binding [12]. However, to positively demonstrate that a TWEAK-induced cellular response is definitely mediated by Fn14, blocking experiments with anti-Fn14 antibodies or Fn14 knockdown is recommended.
2. The TWEAK encoding DNA fragment used for cloning has to cover the complete C-terminal TNF homology domain of the molecule, which mediates Fn14 binding and homotrimerization of TWEAK, and must further start N-terminally to the TWEAK furin processing site to avoid unwanted cleavage of the recombinant protein.
3. TWEAK is highly conserved, thus there is no evidence for species specificity and recombinant variants of human TWEAK can be straightforwardly used for analysis of murine cell lines and mice models.
4. Similar to cell surface antigen-bound IgG1, Fc-Flag-TWEAK is able to trigger ADCC upon binding to Fn14. It is therefore

highly recommended to use for analysis of Fcγ receptor (FcγR)-expressing cells or for in vivo studies Fc-Flag-TWEAK variants with mutations in the Fc part that prevent binding to ADCC-inducing FcγRs without affecting the dimerization capability of the Fc domain.

5. A protocol describing the generation of stable HEK transfectants for large-scale production can be adopted from reference [14].
6. If the anti-Flag agarose will be also used to purify other proteins/TWEAK variants, clean it by acid elution with glycine (0.1 M glycine–HCl, pH 3.5). To protect the anti-Flag agarose matrix, acid elution should not last more than 10 min. Afterwards the column should be immediately equilibrated to a neutral pH.
7. A significant fraction of scFvs has an intrinsic tendency to auto-aggregate. If such scFvs were used for the generation of scFv-Flag-TWEAK fusion proteins this might result in spontaneous oligomerization of the molecule. Like Fc-Flag-TWEAK and oligomerized soluble Flag-TWEAK trimers, the latter might then already mimic membrane TWEAK without cell surface antigen-anchoring. The presence of aggregates in a scFv-Flag-TWEAK preparation must therefore be carefully checked by gel filtration and functional assays. Often there is only a minor fraction of aggregated molecule species in scFv-Flag-TNF ligand fusion protein preparations which can be removed by preparative gel filtration. Be aware that 10 % of aggregates are already sufficient to diminish the difference in the ED50 values of cell surface antigen-bound (or oligomerized) and unanchored (or non-oligomerized) species of TWEAK trimers for activation of the classical NFκB pathway tenfold.
8. Protocol is not suitable for cell lines where Flag-TWEAK per se already induces significant cell death, such as Kym-1 or SKOV3 cells. In such cases the protocol can be adopted to analyze TWEAK-induced cell death simply by omitting TNF. Kym-1 and SKOV3 are already sensitive for TNF- and TWEAK-induced cell death in the absence of CHX so that the latter is also dispensable.
9. To ascertain beyond doubt activation of the alternative NFκB pathway by analyzing p100 to p52 processing, it is not sufficient to demonstrate an increase in p52. P100 is substrate of the classical NFκB pathway, activation of the latter might therefore result in increased expression of the alternative NFκB pathway "substrate" p100 which results in more p52 without an increase in the activity of the alternative NFκB pathway. Misinterpretations related to upregulation of p100 can be prevented by determining the ratio of p100 to p52, analysis of NIK accumulation, and/or by doing experiments in the presence of an IKK2-specifc inhibitor such as TPCA-1.

10. Instead of a pair of cells with and without expression of the cell surface-antigen targeted by the scFv-Flag-TWEAK fusion protein, one can also use a cell surface antigen-positive cell line and perform the assay in presence and absence of a high concentration of the scFv domain-corresponding antibody as a competitor.
11. The analysis of Fn14-mediated effects in coculture experiments with TWEAK expressing stimulator/effector cells and Fn14-expressing responder cells is facilitated by use of cells of different species. Coculture experiments with TWEAK expressing murine transfectants and Fn14-positive target cells allow straightforward detection of target genes of Fn14 signaling when antibodies were used that discriminate between the human and murine version of the antigen.
12. The concentrations of soluble TWEAK accumulating in the supernatants of cells transfected with wild-type full-length TWEAK can reach several ng/mL, a range where soluble TWEAK might already trigger significant cellular responses. In assays using membrane TWEAK expressing transfectants, the latter should be washed prior to the experiment or, alternatively, a mutated variant of TWEAK with a destroyed furin cleavage side should be used [15].
13. It is important to realize that Fn14-stimulating activity in supernatants of membrane TWEAK expressing cells not necessarily implies that this is triggered by soluble TWEAK trimers as membrane TWEAK-containing exosomes might also contribute to Fn14 stimulation. Supernatants must therefore be carefully checked for the presence of membrane TWEAK. If relevant, the latter could be removed by ultracentrifugation.

References

1. Burkly LC, Michaelson JS, Zheng TS (2011) TWEAK/Fn14 pathway: an immunological switch for shaping tissue responses. Immunol Rev 244:99–114
2. Winkles JA (2008) The TWEAK-Fn14 cytokine-receptor axis: discovery, biology and therapeutic targeting. Nat Rev Drug Discov 7: 411–425
3. Grell M, Douni E, Wajant H, Löhden M, Clauss M, Maxeiner B et al (1995) The transmembrane form of tumor necrosis factor is the prime activating ligand of the 80 kDa tumor necrosis factor receptor. Cell 83:793–802
4. Schneider P, Holler N, Bodmer JL, Hahne M, Frei K, Fontana A et al (1998) Conversion of membrane-bound Fas(CD95) ligand to its soluble form is associated with downregulation of its proapoptotic activity and loss of liver toxicity. J Exp Med 187:1205–1213
5. Wajant H, Moosmayer D, Wüest T, Bartke T, Gerlach E, Schönherr U et al (2001) Differential activation of TRAIL-R1 and -2 by soluble and membrane TRAIL allows selective surface antigen-directed activation of TRAIL-R2 by a soluble TRAIL derivative. Oncogene 20:4101–4106
6. Müller N, Wyzgol A, Münkel S, Pfizenmaier K, Wajant H (2008) Activity of soluble OX40 ligand is enhanced by oligomerization and cell surface immobilization. FEBS J 275:2296–22304
7. Wyzgol A, Müller N, Fick A, Munkel S, Grigoleit GU, Pfizenmaier K et al (2009) Trimer stabilization, oligomerization, and antibody-mediated cell surface immobilization

improve the activity of soluble trimers of CD27L, CD40L, 41BBL, and glucocorticoid-induced TNF receptor ligand. J Immunol 183: 1851–1861

8. Bossen C, Cachero TG, Tardivel A, Ingold K, Willen L, Dobles M et al (2008) TACI, unlike BAFF-R, is solely activated by oligomeric BAFF and APRIL to support survival of activated B cells and plasmablasts. Blood 111:1004–1012
9. Schneider P, Schwenzer R, Haas E, Mühlenbeck F, Schubert G, Scheurich P et al (1999) TWEAK can induce cell death via endogenous TNF and TNF receptor 1. Eur J Immunol 29: 1785–1792
10. Wicovsky A, Salzmann S, Roos C, Ehrenschwender M, Rosenthal T, Siegmund D et al (2009) TNF-like weak inducer of apoptosis inhibits proinflammatory TNF receptor-1 signaling. Cell Death Differ 16:1445–1459
11. Roos C, Wicovsky A, Müller N, Salzmann S, Rosenthal T, Kalthoff H et al (2010) Soluble and transmembrane TNF-like weak inducer of apoptosis differentially activate the classical and noncanonical NF-kappa B pathway. J Immunol 185:1593–1605
12. Fick A, Lang I, Schäfer V, Seher A, Trebing J, Weisenberger D et al (2012) Studies of binding of tumor necrosis factor (TNF)-like weak inducer of apoptosis (TWEAK) to fibroblast growth factor inducible 14 (Fn14). J Biol Chem 287:484–495
13. Bover LC, Cardó-Vila M, Kuniyasu A, Sun J, Rangel R, Takeya M et al (2007) A previously unrecognized protein-protein interaction between TWEAK and CD163: potential biological implications. J Immunol 178:8183–8194
14. Fick A, Wyzgol A, Wajant H (2012) Production, purification, and characterization of scFv TNF ligand fusion proteins. Methods Mol Biol 907:597–609
15. Brown SA, Ghosh A, Winkles JA (2010) Full-length, membrane-anchored TWEAK can function as a juxtacrine signaling molecule and activate the NF-kappaB pathway. J Biol Chem 285:17432–17441

Chapter 5

Regulation of Human Dendritic Cell Functions by Natural Anti-CD40 Antibodies

Sri Ramulu Elluru, Srini V. Kaveri, and Jagadeesh Bayry

Abstract

Dendritic cells (DCs) are professional antigen presenting cells that play a pivotal role in the initiation of immune responses. DCs ingest antigens and then present these antigens to T cells to initiate T cell activation and polarization. DCs receive signals both from environment and from endogenous molecules. DCs in the immune system constantly interact with immunoglobulins (or antibodies) and a substantial amount of these immunoglobulins are natural. We found that natural antibodies have a key role in regulating the DC functions and that CD40-reactive natural antibodies constitute one of the endogenous molecules that provide maturation-associated signals to DCs in physiology. In this chapter, we describe the isolation of anti-CD40 natural antibodies from pooled normal immunoglobulin preparations (intravenous immunoglobulin, IVIg) and their biological effects on human DC maturation and functions.

Key words CD40, Natural antibodies, Dendritic cells, T cells, Maturation, Affinity chromatography purification, Intravenous immunoglobulin

1 Introduction

Dendritic cells (DCs) are professional antigen presenting cells (APCs) and are critical for both immune tolerance and in the pathogenesis of autoimmune and inflammatory conditions. DCs at immature state are poor stimulators of T cells, express low level of co-stimulatory molecules, and secrete minimal amounts of inflammatory cytokines and chemokines. However, immature DCs express diverse sensing molecules that include pattern-recognition receptors and receptors for inflammatory cytokines. Upon receiving activation signaling, either from pathogens, necrotic cells, or inflammatory cytokines, DCs undergo maturation and activation, migrate to secondary lymphoid tissues, and present antigenic peptides in the context of MHC molecules to naïve T cells to initiate immune responses. Therefore, dissection of signals that regulate functions of DCs is highly important to understand the molecular

Jagadeesh Bayry (ed.), *The TNF Superfamily: Methods and Protocols*, Methods in Molecular Biology, vol. 1155,
DOI 10.1007/978-1-4939-0669-7_5, © Springer Science+Business Media New York 2014

mechanisms of immune tolerance and pathogenesis of autoimmune and inflammatory disorders [1–3].

The role of immunoglobulins (or antibodies), the product of B cells, is not restricted to their pathogenic functions as in autoimmune diseases, rather they have a diverse immunoregulatory functions. These functions include neutralization of pathogens and their products, opsonization of antigens, removal of aged and transformed cells, antibody-dependent cellular cytotoxicity, and protecting the tissues from inflammation [4–10]. In addition, immunoglobulins also regulate the functions of immune cells both via Fc-dependent interaction with Fc receptors and F(ab)-dependent interaction with non-Fc receptors [4, 11–13]. Of note, significant quantities of immunoglobulins are natural. Natural antibodies are germ-line encoded and produced in the absence of deliberate immunization [14–16]. Several immunoregulatory functions have been proposed for natural antibodies [5, 6, 8, 17, 18].

We report that natural antibodies play a key role in the regulation of DC functions. In fact, DCs from patients with primary immunodeficiencies such as X-linked agammaglobulinemia and common variable immunodeficiency displayed defective differentiation and that ex vivo reconstitution of patients' plasma with normal circulating immunoglobulins (intravenous immunoglobulin, IVIg) partially rescued the differentiation process of patients' DCs [19–21]. As interaction between CD40 on DCs and CD40L on T cells is involved in a wide range of immunological cross-talks including promotion of differentiation and maturation of DCs [22–26], we surmised that natural antibody repertoire contains antibodies to CD40 that provide maturation-associated signals to DCs in physiology. CD40 is a type I transmembrane glycoprotein of 48–50 kDa, belonging to the TNF-receptor family. In this chapter, we describe the isolation of anti-CD40 natural antibodies from pooled normal immunoglobulin preparations (IVIg) and their biological effects on human DC maturation and functions.

2 Materials

2.1 General Equipments

1. Biological safety cabinet for cell culture.
2. Inverted microscope with ×10 and ×20 objectives.
3. A 37 °C incubator with humidity and gas control to maintain >95 % humidity and an atmosphere of 5 % CO_2 in air.
4. Low-speed centrifuge.
5. Other equipments and supplies: sterile 6-, 24-, and 96-well plates, tissue culture flasks (T25, T75), plastic pipettes (1, 5,10, 25 mL), micropipettes (20, 200, 1,000 μL), tips (20, 200, 1,000 μL), multichannel pipettes, 15 and 50 mL conical

tubes, sintered glass filter, rotator, horizontal shaker, affinity columns, affinity column holding stands, hemocytometer, water bath, spectrophotometer, refrigerator, deep freezer.

2.2 Purification of Natural Anti-CD40 Antibodies

1. CNBr-activated Sepharose 4B beads (Pharmacia Biotech). 1 g gives 3 mL bed volumes.
2. CD40 peptide: ^{78}HQHKYCDPNLGLRV^{91} (*see* **Note 1**).
3. Peptide coupling buffer: 0.1 M $NaHCO_3$, pH 8.3, 0.5 M NaCl.
4. Quenching buffer (Tris–EDTA–NaCl): 0.5 M NaCl, 0.1 M Tris–HCl, pH 8.2, 1 mM EDTA.
5. Affinity column and Sepharose 4B beads wash buffer: PBS, 0.01 % NaN_3.
6. Intravenous immunoglobulin (IVIg) (*see* **Note 2**).
7. Antibody elution buffer: 0.2 M glycine–HCl, pH 2.8.
8. Neutralization buffer: 3 M Tris–HCl.
9. PBS: 137 mM NaCl, 2.7 mM KCl, 8 mM Na_2HPO_4, 2 mM KH_2PO_4. Dilute 10× PBS to 1× PBS.
10. Centriprep® centrifugal filter units (Millipore).

2.3 ELISA

1. ELISA plate reader.
2. ELISA 96-well plate.
3. Blocking buffer: PBS, 1 % bovine serum albumin (BSA).
4. Washing buffer: PBS, 0.1 % Tween 20.
5. Dilution Buffer: PBS, 0.1 % Tween 20, 0.1 % BSA.
6. Detection antibody: polyclonal mouse anti-human IgG-HRP.
7. Substrate: *o*-Phenylenediamine dihydrochloride (OPD, Sigma).
8. Stop solution: 2 N HCl.

2.4 Generation of Monocyte-Derived Dendritic Cells

1. RPMI 1640 medium.
2. Complete RPMI 1640 medium: RPMI medium with 10 % fetal calf serum (FCS), 100 U/mL penicillin, 100 μg/mL streptomycin.
3. CD14 MicroBeads (Miltenyi Biotec) (*see* **Note 3**).
4. MACS buffer: Degassed PBS, 1 % BSA. Always maintain at 4 °C.
5. MACS LS and MS columns (Miltenyi Biotec).
6. MACS magnetic stand and magnet (Miltenyi Biotec).
7. Ficoll-Paque ($p = 1.077$ g/mL).
8. Recombinant human GM-CSF and IL-4.
9. Heparinized blood or buffy bags from healthy donors.

3 Methods

3.1 Coupling of CD40 Peptide to CNBr-Activated Sepharose Beads

1. Soak 1 g of CNBr-activated Sepharose 4B in 50 mL of 0.1 M HCl for 15–20 min.
2. Filter the Sepharose beads on a sintered glass filter and wash the cake with a few hundred milliliter of 0.1 M HCl.
3. Transfer the Sepharose cake into 50 mL tube covered with foil.
4. Dissolve the CD40 peptide (1 mg/mL) in coupling buffer.
5. Mix the peptide with the cake in 50 mL tube in coupling buffer and left overnight at 4 °C with slow rotation.
6. Centrifuge the tube slowly at 50 × *g* 4 °C for few min.
7. Transfer the whole gel into sintered glass filter and collect the flow through for checking the binding efficiency of the peptide.
8. Wash the gel with quenching buffer, resuspend in quenching buffer, and rotate for 2 h at 4 °C.
9. Transfer the gel onto a sintered glass filter and wash with three volumes of quenching buffer.
10. Wash the gel with four volumes of 0.5 M NaCl.
11. Wash the gel with two volumes of quenching buffer.
12. Wash the gel with one volumes of 0.2 M glycine–HCl.
13. Wash the gel with two volumes of quenching buffer.
14. Repeat the process as in **steps 10–13**.
15. Wash the gel with two volumes of affinity column and Sepharose 4B beads wash buffer, and collect the matrix into a 50 mL tube and resuspend in affinity column and Sepharose 4B beads wash buffer and store at 4 °C.

3.2 Isolation of Anti-CD40 Natural Antibodies

1. Transfer the CD40 peptide-coupled Sepharose beads into the affinity column.
2. Incubate 10 mg of IVIg/mL of affinity matrix for 3 h at room temperature or overnight at 4 °C.
3. Wash the column with affinity column and Sepharose 4B beads wash buffer and collect it as flow through (FT).
4. Then proceed to elution with antibody elution buffer. Make sure that you put just 1 mL of elution buffer and collect the eluate in a 15 mL tube.
5. Add another 5 mL of antibody elution buffer to the column and add 200 μL of neutralization buffer to eluate in the tube.
6. Dialyze the eluate against PBS without azide, at least twice at 4 °C (*see* **Note 4**).
7. Concentrate the anti-CD40 natural antibodies by using Centriprep® centrifugal filter units as per the instructions, sterile filter through syringe-fit 0.22 μm filters and store at −20 °C.

8. Wash the affinity column with another 20–40 mL of antibody elution buffer to dissociate any other IgG bound to the column.
9. Wash the affinity column with at least 50 mL of affinity column and Sepharose 4B beads wash buffer before storing away the column at 4 °C.

3.3 Binding Studies of Anti-CD40 Natural Antibodies to CD40 Peptide by ELISA

1. Coat 96-well ELISA plate with 3 μg/well/100 μL of the CD40 peptide for 3 h at 37 °C.
2. Remove the contents, wash with 200 μL of PBS-0.1 % Tween and block the plates with 200 μL of PBS-1 % BSA for 1 h at 37 °C.
3. Wash the plate three times with PBS-0.1 % Tween and then add in triplicates serial dilutions (4–512 μg/mL) of IVIg, purified anti-CD40 natural antibodies and flow through fraction that was depleted of anti-CD40 natural antibodies at 100 μL volume and incubate for 2 h at 37 °C.
4. Aspirate and wash the plates three times and detect the bound IgG with secondary antibody (mouse anti-human IgG-HRP, 1 μg/mL). Add 100 μL of detection antibody to each well and incubate at room temperature for 2 h.
5. Aspirate and wash the wells three times, add 100 μL of OPD as its substrate and incubate in the dark.
6. After the development of the color, stop the reaction by adding 100 μL of stop solution to each well and gently mix the solution.
7. Record the optical density at 492 nm using plate reader.

3.4 Generation of Human Monocyte-Derived DCs (MO-DCs)

1. Dilute the blood at least two times with RPMI 1640 medium.
2. In a 50 mL Falcon tube, layer the blood gently on the layer of Ficoll-hypaque using a 25 mL pipette. For every 15 mL of Ficoll-hypaque, lay 30 mL of blood. Take extreme care to avoid the mixing of these two layers.
3. Spin the tubes at 400 × *g* for 30 min in a swinging bucket rotor at room temperature without brake. Please check that the brake is off or the acceleration and deceleration of the centrifuge is zero. This ensures that the two layers do not mix.
4. After the spin, using a 10 mL pipette, gently aspirate the intermediate, translucent and white layer containing peripheral blood mononuclear cells (PBMC) into the new tube. Try to take as much as possible. Add large volumes of RPMI 1640 medium to the tube to reduce the toxic effects of Ficoll-hypaque.
5. Wash the PBMC using RPMI 1640 medium at 300 × *g* for 5–10 min at 4 °C. Count the cells and prepare for the next step.

6. Add few milliliter of MACS buffer to PBMC and wash the cells at 300 × *g* for 5 min at 4 °C.
7. Resuspend the cell pellet in 80 μL of MACS buffer per 10^7 PBMC and add 20 μL of CD14 MicroBeads. Mix the cell suspension gently.
8. Incubate the cells at 4 °C (refrigerator) for 15 min (*see* **Note 5**).
9. After 15 min, wash cells by adding 1–2 mL of MACS buffer for ten million cells. Centrifuge the cells at 300 × *g* for 10 min at 4 °C.
10. Resuspend the cell pellet in a volume of 500 μL for up to 100 million cells. This minimum volume needs to be maintained.
11. Prepare the MS/LS column by adding recommended volume of buffer. Then apply cell suspension to the column. Collect the unlabelled cells in the flow through.
12. Then wash the column thrice with recommended volume of buffer (Miltenyi Biotec data sheet). Add new buffer only when the column reservoir is empty.
13. Remove column from the separator and place it on a suitable collection tube. Add suitable volume of buffer to the column as per instructions provided in Miltenyi Biotec data sheet. Quickly flush the contents using the plunger given along with the column.
14. Wash the purified $CD14^+$ monocytes in complete RPMI medium. Count the cells and resuspend the cells in a volume of one million cells/mL in complete RPMI medium containing 1,000 U/mL of recombinant human GM-CSF and 500 U/mL of recombinant human IL-4.
15. Incubate the monocytes, in recommended volume in the appropriate flasks in humidified 5 % CO_2 incubator at 37 °C (*see* **Note 6**).
16. DCs are differentiated in about 5–7 days (*see* **Note** 7).

3.5 Treatment of MO-DCs with Natural Anti-CD40 Natural Antibodies

1. Wash the differentiated MO-DCs twice with RPMI 1640 medium. Count the cells and resuspend in appropriate volume of complete RPMI 1640 medium (0.5 million DCs/mL) with GM-CSF and IL-4.
2. Plate DCs in 12-well tissue culture plate at 0.5 million DCs/mL/well.
3. Add natural anti-CD40 antibodies at various concentrations (25–100 μg/well).
4. Incubate the MO-DCs with the antibodies for 48 h in humidified 5 % CO_2 incubator at 37 °C.
5. Pellet the DCs at 300 × *g* for 5 min, collect the supernatants, and store at −20 °C for the analysis of cytokines and chemokines. Proceed with the analysis of phenotype of DCs by flow cytometry by using fluorochrome-conjugated monoclonal antibodies (*see* **Note 8**).

4 Notes

1. Selection of peptide sequence is important for the isolation of functionally relevant natural antibodies from circulating immunoglobulins. The peptide sequence of CD40 used by us comprises three of the seven residues critical for the binding of CD40 to its ligand (Y^{82}, D^{84}, N^{86}) [27].
2. IVIg is a pooled normal IgG preparation obtained from plasma of several thousand healthy blood donors [7, 9, 10]. IVIg represents a privileged source of natural antibodies. The diversity of antibody repertoire as a result of large donor pool ensures antibody specificities to a wide spectrum of self and foreign antigens. IVIg is commercially available. If the product is lyophilized, dissolve the product with PBS/RPMI 1640 and dialyze at least three times against large volume of PBS to remove stabilizing agents. If IVIg is in liquid form, then proceed for the dialysis directly.
3. Previously monocytes were isolated by plate adherence method. However, currently, this method is replaced by MicroBead-based techniques and automated cell sorters.
4. Avoid sodium azide during dialysis of isolated antibodies against PBS. Sodium azide is toxic to cells.
5. It is recommended to incubate the cells at 4 °C (refrigerator) for labeling with MicroBeads and not on ice. Incubation in ice reduces the labeling efficiency.
6. Human DCs can also be isolated from the circulation. However, circulating DCs are low in number (less than 1 % of PBMC) and are heterogeneous. Because of their higher numbers (4–15 % of PBMC), monocytes are commonly used to differentiate human DCs with uniform features.
7. For the functional assays, it is ideal to use 5–7 day-old differentiated immature DCs. After 7 days, DCs gradually loose their potency.
8. Purity of the monocytes or the DCs can be analyzed by FACS using CD14/CD16 (monocytes/macrophages) or CD1a, DC-SIGN (DCs) antibodies. CD80, CD40, CD86, DC-LAMP, HLA-DR can be used as markers to check the effect of anti-CD40 natural antibodies on DC maturation.

Acknowledgements

This work was supported by the Institut National de la Santé et de la Recherche Médicale (INSERM), Centre National de la Recherche Scientifique (CNRS), Université Pierre et Marie Curie, and Université Paris Descartes

References

1. Banchereau J, Steinman RM (1998) Dendritic cells and the control of immunity. Nature 392:245–252
2. Steinman RM, Banchereau J (2007) Taking dendritic cells into medicine. Nature 449:419–426
3. Banchereau J, Briere F, Caux C, Davoust J, Lebecque S, Liu YJ et al (2000) Immunobiology of dendritic cells. Annu Rev Immunol 18: 767–811
4. Schroeder HW Jr, Cavacini L (2010) Structure and function of immunoglobulins. J Allergy Clin Immunol 125:S41–S52
5. Ehrenstein MR, Notley CA (2010) The importance of natural IgM: scavenger, protector and regulator. Nat Rev Immunol 10:778–786
6. Kaveri SV, Silverman GJ, Bayry J (2012) Natural IgM in immune equilibrium and harnessing their therapeutic potential. J Immunol 188:939–945
7. Kazatchkine MD, Kaveri SV (2001) Immunomodulation of autoimmune and inflammatory diseases with intravenous immune globulin. N Engl J Med 345:747–755
8. Ochsenbein AF, Fehr T, Lutz C, Suter M, Brombacher F, Hengartner H et al (1999) Control of early viral and bacterial distribution and disease by natural antibodies. Science 286:2156–2159
9. Bayry J, Negi VS, Kaveri SV (2011) Intravenous immunoglobulin therapy in rheumatic diseases. Nat Rev Rheumatol 7:349–359
10. Tha-In T, Bayry J, Metselaar HJ, Kaveri SV, Kwekkeboom J (2008) Modulation of the cellular immune system by intravenous immunoglobulin. Trends Immunol 29:608–615
11. Nimmerjahn F, Ravetch JV (2008) Fcgamma receptors as regulators of immune responses. Nat Rev Immunol 8:34–47
12. Bayry J, Lacroix-Desmazes S, Kazatchkine MD, Hermine O, Tough DF, Kaveri SV (2005) Modulation of dendritic cell maturation and function by B lymphocytes. J Immunol 175:15–20
13. Elluru SR, Vani J, Delignat S, Bloch MF, Lacroix-Desmazes S, Kazatchkine MD et al (2008) Modulation of human dendritic cell maturation and function by natural IgG antibodies. Autoimmun Rev 7:487–490
14. Avrameas S (1991) Natural autoantibodies: from 'horror autotoxicus' to 'gnothi seauton'. Immunol Today 12:154–159
15. Coutinho A, Kazatchkine MD, Avrameas S (1995) Natural autoantibodies. Curr Opin Immunol 7:812–818
16. Wardemann H, Yurasov S, Schaefer A, Young JW, Meffre E, Nussenzweig MC (2003) Predominant autoantibody production by early human B cell precursors. Science 301: 1374–1377
17. Kohler H, Bayry J, Nicoletti A, Kaveri SV (2003) Natural autoantibodies as tools to predict the outcome of immune response? Scand J Immunol 58:285–289
18. Zhou ZH, Wild T, Xiong Y, Sylvers P, Zhang Y, Zhang L et al (2013) Polyreactive antibodies plus complement enhance the phagocytosis of cells made apoptotic by UV-light or HIV. Sci Rep 3:2271
19. Bayry J, Lacroix-Desmazes S, Donkova-Petrini V, Carbonneil C, Misra N, Lepelletier Y et al (2004) Natural antibodies sustain differentiation and maturation of human dendritic cells. Proc Natl Acad Sci U S A 101:14210–14215
20. Bayry J, Lacroix-Desmazes S, Hermine O, Oksenhendler E, Kazatchkine MD, Kaveri SV (2005) Amelioration of differentiation of dendritic cells from CVID patients by intravenous immunoglobulin. Am J Med 118: 1439–1440
21. Bayry J, Lacroix-Desmazes S, Kazatchkine MD, Galicier L, Lepelletier Y, Webster D et al (2004) Common variable immunodeficiency is associated with defective functions of dendritic cells. Blood 104:2441–2443
22. Schoenberger SP, Toes RE, van der Voort EI, Offringa R, Melief CJ (1998) T-cell help for cytotoxic T lymphocytes is mediated by CD40–CD40L interactions. Nature 393:480–483
23. Caux C, Massacrier C, Vanbervliet B, Dubois B, Van Kooten C, Durand I et al (1994) Activation of human dendritic cells through CD40 cross-linking. J Exp Med 180:1263–1272
24. Cella M, Scheidegger D, Palmer-Lehmann K, Lane P, Lanzavecchia A, Alber G (1996) Ligation of CD40 on dendritic cells triggers production of high levels of interleukin-12 and enhances T cell stimulatory capacity: T-T help via APC activation. J Exp Med 184:747–752
25. O'Sullivan BJ, Thomas R (2002) CD40 ligation conditions dendritic cell antigen-presenting function through sustained activation of NF-kB. J Immunol 168:5491–5498
26. Fujii S, Liu K, Smith C, Bonito AJ, Steinman RM (2004) The linkage of innate to adaptive immunity via maturing dendritic cells in vivo requires CD40 ligation in addition to antigen presentation and CD80/86 costimulation. J Exp Med 199:1607–1618
27. Bajorath J, Chalupny NJ, Marken JS, Siadak AW, Skonier J, Gordon M et al (1995) Identification of residues on CD40 and its ligand which are critical for the receptor-ligand interaction. Biochemistry 34:1833–1844

Chapter 6

β-Actin in the Signaling of Transmembrane TNF-α-Mediated Cytotoxicity

Hui Chen, Yan Leng, and Zhuoya Li

Abstract

To study the role of β-actin in the signaling of transmembrane TNF-α (tmTNF-α)-mediated cytotoxicity, we mainly used bioassay and apoptosis assay, detection of the visual changes in β-actin and intracellular translocation of signal molecules in response to tmTNF-α, and analysis of the signal molecules coupled or uncoupled with TNFR2 complex. These protocols might also be used to investigate the signaling events mediated by other transmembrane cytokines or members of TNF superfamily, which have transmembrane and soluble forms, as well as the involvement of actin cytoskeleton in signal transduction pathways.

Key words Transmembrane TNF-α, Actin cytoskeleton, Cytotoxicity, Confocal microscopy, IP-Western blot

1 Introduction

Tumor necrosis factor-α is synthesized primarily as a 26-kDa transmembrane molecule (tmTNF-α). tmTNF-α is then cleaved mainly by a metalloproteinase, the TNF-alpha converting enzyme (TACE), releasing a 17-kDa soluble cytokine (sTNF-α) [1]. Both forms of TNF-α display bioactivities via TNFR1 and TNFR2. TNFR1 mediates the principal functions of sTNF-α. Besides initiating TNFR1 signaling, tmTNF-α is also a primary ligand for TNFR2 [2, 3]. As their names imply, both forms of TNF-α exhibit cytotoxicity against tumor cells, and tmTNF-α can even kill sTNF-resistant neoplastic cells [4, 5]. To study the tmTNF-α-mediated cytotoxicity and signaling, we are faced with three problems. First, TNF-α is present in both forms in vitro as well as in vivo and tmTNF-α can be constantly processed into sTNF-α. Therefore, in order to monitor the function of tmTNF-α, its amount expressed on the cell surface should be sufficient and at the same time the effect of sTNF-α should be excluded. Second, tmTNF-α is a membrane-bound molecule and exerts its activities in a cell–cell contact manner. It is therefore clear that other membrane

Jagadeesh Bayry (ed.), *The TNF Superfamily: Methods and Protocols*, Methods in Molecular Biology, vol. 1155, DOI 10.1007/978-1-4939-0669-7_6, © Springer Science+Business Media New York 2014

molecules on the cell surface may affect the function of tmTNF-α. For example, cytotoxic membrane molecules such as FasL and TRAIL should be absent or excluded in the experiment system without interfering the adhesion molecules-mediated contact between tmTNF-α-expressing cells and target cells [6]. Third, it should be pondered seriously over how the target cells in the coculture could be separated from the tmTNF-α-expressing cells for further study. In this chapter, we discuss the ways of dealing with these problems.

The actin cytoskeleton is a dynamic structure integrally involved in the coupling of extracellular stimuli to cell activation and concomitant changes in morphology, motility, endocytosis, and many other fundamental cellular processes. It has been reported that actin cytoskeleton plays a regulating role in the signal transduction pathways involving antigen receptors in T and B cells [7, 8], the mitogen-activated protein kinase (MAPK) pathways [9, 10], and the NF-κB pathway [11–13]. The latter two pathways can also be activated by cognate ligation of TNFR. In addition, the dynamic structure of the actin cytoskeleton has been demonstrated to play a crucial role in apoptosis and necrosis. Increased turnover of F-actin is important for longevity of the cell, while decreased actin turnover results in apoptosis-like cell death [14, 15]. Currently, two main approaches are used to study the involvement of actin cytoskeleton in the signal transduction pathways: (1) Modulation of the expression or the activity of endogenous actin dynamics-regulating proteins like the small RhoGTPases, which mediate actin polymerization and actin cytoskeleton remodeling in response to extracellular stimuli. (2) Usage of actin-targeting natural compounds, which directly interact with actin-G or actin-F and interfere with actin dynamics. However, anti-actin drugs are more frequently used because they target more specifically the actin dynamics than small RhoGTPases [16, 17]. In our study, we used F-actin-destabilizing compounds cytochalasin D (CytD) and latrunculin A (LatA) to disturb actin polymerization so as to observe by confocal microscopy, the visual changes between actin and intracellular translocation of signal molecules in response to tmTNF-α. At the same time, immunoprecipitation analysis was adopted to detect the signal molecules coupled with or uncoupled from TNFR2 signal complex. As expected, the results showed that actin cytoskeleton is actually involved in TNFR2 signaling for tmTNF-α-mediated cytotoxicity [18]. This chapter describes the protocol involved in tmTNF-α-mediated cytotoxicity, detection of actin, and the signal transduction in response to tmTNF-α with or without anti-actin drug. We hope that this protocol could be of helpful to study other transmembrane cytokines or TNF superfamily members (many of them do have transmembrane and soluble forms) as well as the involvement of actin in signal transduction pathways.

2 Materials

2.1 Cell Culture and Stimulus

1. Complete cell culture medium: RPMI-1640 medium, 10 % heat-inactivated, pyrogen-free fetal calf serum (FCS), 1.0 mM sodium pyruvate, 2.0 mM L-glutamine, 100 U/mL penicillin, 100 μg/mL streptomycin, 5×10^{-5} M 2-ME.
2. tmTNF-α acquisition.
   (a) PBS (phosphate-buffered saline): 120 mM NaCl, 1.2 mM NaH_2PO_4, 2.8 mM KCl, 8.8 mM Na_2HPO_4, pH 7.4.
   (b) 1 % Paraformaldehyde: 0.1 g paraformaldehyde in 10 mL of PBS.
   (c) Acid glycine buffer: 50 mM glycine–HCl, pH 3.0, 150 mM NaCl.

2.2 tmTNF-α Cytotoxicity

1. The choice of effector cells and target cells: Besides consideration of sensitivity of target cells to TNF-α, we usually choose adherent cell that expresses tmTNF-α as an effector cell and suspension cell as a target cell that is easily separated from the coculture for further study (*see* **Note 1**).
2. Source of target cells: A suspension promyelocytic leukemia cell line HL-60.
3. Source of tmTNF-α (effector cells): An adherent fibroblast cell line NIH3T3 as an effector cell, which does not express FasL, TRAIL, and other cytotoxic molecules. NIH3T3 cell line was stably transfected with an empty retrovirus plasmid pLXSN as a control, or pLW-TNF-α, the plasmid inserted with human wtTNF-α cDNA at the *Hpa*I and *Xho*I cloning site, in our laboratory [19] (*see* **Note 2**).
4. 96-Well microtiter plates (Biomat, Italy).
5. Cytochalasin D (CytD) (Invitrogen, USA) (*see* **Note 3**).
6. MTT assay kit (Sigma, USA).
7. Microplate reader (Tecan, Austria).

2.3 tmTNF-α Apoptosis Detection

1. 6-Well microtiter plates (Biomat).
2. LatA (Invitrogen).
3. Annexin V-FITC and Propidium iodide (PI) kit (BD Biosciences, USA).
4. A FACSCalibur 440E flow cytometer (Becton Dickinson, USA).

2.4 Confocal Microscopy

1. Antibodies specific to β-actin and signal molecules TRAF2 and RIP1 (Santa Cruz, USA).
2. The matched fluorescence-conjugated secondary antibodies (Jackson Biotech, USA).

3. PI (Sigma).
4. A confocal microscope FU5000 (Olympus, Japan).

2.5 Actin Polymerization Assay

1. Rhodamine/FITC-phalloidin (Molecular Probes®, Invitrogen).
2. A FACSCalibur 440E flow cytometer.

2.6 Cytoplasmic, Nuclear, and Total Protein Preparation

1. Lysis buffer A: 10 mM HEPES, pH 7.8, 10 mM KCl, 0.1 mM EDTA, 1 mM DTT, 0.5 mM PMSF, 5 μg/mL aprotinin, 5 μg/mL leupeptin (Sigma). Store at 4 °C (*see* **Note 4**).
2. Lysis buffer C: 50 mM HEPES, pH 7.8, 0.42 M KCl, 0.1 mM EDTA, 1 mM DTT, 0.5 mM PMSF, 5 μg/mL aprotinin, 5 μg/mL leupeptin, 5 mM $MgCl_2$, 20 % glycerin. Store at 4 °C.
3. Total protein lysis buffer: 30 mM Tris–HCl, pH 7.5, 150 mM NaCl, 1 % NP-40, 10 % glycerol, 0.5 mM PMSF, 5 μg/mL aprotinin, 5 μg/mL leupeptin. Store at 4 °C.

2.7 Immuno-precipitation (IP)

1. IP lysis buffer: 30 mM Tris–HCl, pH 7.5, 150 mM NaCl, 1 % NP-40, 0.5 mM EDTA, 10 % glycerin, 1 mM PMSF, 5 μg/mL aprotinin, 5 μg/mL leupeptin. Store at 4 °C.
2. Anti-TNFR2 monoclonal antibody (mAb) (Santa Cruz Biotechnology).
3. Protein A agarose (Santa Cruz Biotechnology).
4. 4× SDS sample buffer: 200 mmol/L Tris–HCl, pH 6.8, 400 mmol/L DTT, 8 % SDS, 0.4 % bromophenol blue, 40 % glycerol.

2.8 Western Blot Analysis

1. 12 % Resolving gel: 2 mL 30 % Acrylamide/bis, 1.6 mL ddH_2O, 1.3 mL 1.5 M Tris–HCl, pH 8.8, 50 μL 10 % SDS, 50 μL 10 % ammonium persulfate, 4 μL TEMED (Sigma) (*see* **Note 5**).
2. 5 % Stacking gel: 0.83 mL 30 % Acrylamide/bis, 3.4 mL ddH_2O, 0.63 mL 1.0 M Tris–HCl, pH 6.8, 50 μL 10 % SDS, 50 μL 10 % ammonium persulfate, 5 μL TEMED.
3. Protein marker (Fermentas, Canada).
4. Running buffer: 0.025 M Tris, pH 8.3, 0.25 M glycine, 0.1 % SDS.
5. Transfer buffer: 0.048 M Tris, 0.039 M glycine, 20 % methanol.
6. Tris-buffered saline (10× TBS): 1.5 M NaCl, 0.1 M Tris–HCl, pH 7.4.
7. TBS-T: TBS containing 0.1 % Tween-20.
8. Antibodies specific to caspase-8, cFLIP, IκB, NF-κB p65 and GAPDH (Santa Cruz Biotechnology).
9. HRP-conjugated anti-mouse IgG, anti-rabbit IgG, and anti-goat IgG antibodies (Jackson Biotech).

10. SDS-PAGE electrophoresis system (BioRad, USA).
11. ECL kit (NEN Life Sciences, USA).
12. Kodak image station 4,000 mm (BD Biosciences Co., USA).

2.9 Activity Analysis of NF-κB by ELISA

1. Lysis buffer: 20 mM HEPES, pH 7.5, 0.35 M NaCl, 20 % glycerol, 1 % NP-40, 1 mM $MgCl_2$, 0.5 mM EDTA, 0.1 mM EGTA, 0.5 mM PMSF, 5 μg/mL aprotinin, 5 μg/mL leupeptin. Store at 4 °C.
2. Two single-stranded oligonucleotide chains: 5′-AGTTGA-GGGGACTTTCCCAGGC-C-(C)34-C-3-bio′, which is biotinylated at the 3′ end, and 5′-GCCTGGGAAAGTCCC-CTCAACT-3′.
3. PBS-T: PBS containing 0.1 % Tween-20.
4. Binding buffer: 4 mM HEPES, pH 7.5, 100 mM KCl, 8 % glycerol, 5 mM DTT, 0.2 % BSA, 40 μg/mL salmon sperm DNA. Store at 4 °C.
5. Peroxidase-conjugated goat anti-mouse IgG.
6. 0.2 % TMB: 10 μg tetramethylbenzidine, 5 mL dehydrated alcohol.
7. Substrate buffer: 486 μL 100 mM citric acid, 514 μL 200 mM Na_2HPO_4.
8. Stop solution: 2 M H_2SO_4.
9. Streptavidin-coated 96-well strip plate (Cayman Chemical).

3 Methods

3.1 tmTNF-α-Mediated Cytotoxicity

3.1.1 Bioassay for tmTNF-α-Mediated Cytotoxicity

1. Culture all cell lines used in this protocol at 37 °C, 5 % CO_2 in complete cell culture medium.
2. Seed 5×10^4/100 μL of target cells (HL-60) per well in a 96-well microtiter plate and incubate with 2 μM of CytD for 2 h at 37 °C 5 % CO_2.
3. Resuspend NIH3T3 cell line stably transfected with either empty plasmid pLXSN (control) or pLW-TNF-α (the plasmid inserted with wtTNF-α) in PBS after detachment by trypsin digestion and wash two times with PBS.
4. Resuspend tmTNF-α-overexpressing NIH3T3 cells in 1 % paraformaldehyde and fix for 30 min at 37 °C with intermittent shaking (*see* **Note 6**).
5. Wash the fixed cells three times with PBS, centrifuge at 1,258 × *g* for 10 min (*see* **Note 7**) and remove supernatant.
6. Incubate the cells with acid glycine buffer for 15 min at room temperature (RT), to remove receptor-bound sTNF-α.

7. Wash cells three times with PBS and remove the supernatant as per the **step 5**.
8. Count the cells using a microscope and adjust appropriate number of NIH3T3 effector cells.
9. Add 5×10^5/100 μL of fixed tmTNF-α-expressing NIH3T3 cells to target cells (at an E:T ratio of 10:1) and incubate for 48 h at 37 °C, 5 % CO_2. Target cells alone, effector cells alone, and CytD alone-treated target cells serve as controls.
10. For the neutralization of TNF-α, treat the same amount of the fixed transfectants with a specific antibody for 1 h at 37 °C, and then wash NIH3T3 cells three times with PBS (1,258 × *g* for 10 min) to remove excess free antibody before addition to the target cells.
11. Add 10 μL of 30 mM glucose–PBS containing 0.5 mg/mL MTT and incubate for 4 h at 37 °C in dark (*see* **Note 8**).
12. Remove the supernatant and add 0.1 mL of DMSO per well to lyse cells.
13. Incubate on an orbital shaker at RT for 10 min in dark.
14. Read the plate at 570 nm in ELISA Reader (background wavelength is 630 nm).
15. tmTNF-α-induced cytotoxicity is calculated by the following formula: where control = target cells alone:

$$\text{Cell death rate } (\%) = (1 - OD_{sample}/OD_{control}) \times 100\ \%.$$

3.1.2 Apoptosis Detection

1. Plate 5×10^5 NIH3T3 transfectants per well in a 6-well microtiter plate and incubate overnight at 37 °C, 5 % CO_2 in complete cell culture medium, allowing cells to adhere to the wells and grow.
2. Fix NIH3T3 transfectants with 1 % paraformaldehyde for 30 min in the plate at 37 °C with intermittent shaking.
3. Wash cells carefully three times with 2 mL of PBS (*see* **Note 9**).
4. Remove sTNF-α bound to TNFR by incubation of the cells with acid glycine buffer for 15 min at RT.
5. Repeat **step 3**.
6. Pretreat HL-60 cells with 2 μM of CytD or 400 ng/mL of LatA for 2 h at 37 °C, 5 % CO_2 (*see* **Note 10**).
7. Add 5×10^4/well pretreated HL-60 target cells to the fixed NIH3T3 transfectants at an E:T ratio 10:1 and incubate for 24 h at 37 °C, 5 % CO_2.
8. Harvest the target cells from the coculture by gently pipetting up and down a few times without affecting adherent effector cells.
9. Wash cells twice with pre-cold PBS, centrifuge at 200 × *g* for 5 min, and remove the supernatant.

10. Double-stain HL-60 cells on ice with Annexin V-FITC and PI in dark, according to the manufacturer's instruction.
11. Analyze apoptosis of target cells on a FACSCalibur 440E flow cytometer.
12. The apoptosis rate induced by tmTNF-α calculated by the following formula:

    Apoptosis rate (%) = the rate of Annexin V-FITC single positive cells + the rate of Annexin V-FITC and PI double-positive cells [20].

3.2 β-Actin Detection

3.2.1 Confocal Microscopy

1. Fix 5×10^5/well NIH3T3 transfectants with 1 % paraformaldehyde in a 6-well microtiter plate (*see* **steps 2** and **3** in Subheading 3.1.2).
2. Remove sTNF-α bound to TNFR on NIH3T3 transfectants (*see* **steps 4** and **5** in Subheading 3.1.2).
3. Pretreat HL-60 cells with 2 μM of CytD for 2 h at 37 °C, 5 % CO_2.
4. Add 4×10^5/well HL-60 cells to NIH3T3 transfectants at an E:T ratio 10:1 and incubate for 25 min at 37 °C, 5 % CO_2.
5. Harvest the target cells from the coculture by gently pipetting up and down a few times without affecting adherent effector cells.
6. Wash HL-60 cells twice with pre-cold PBS, centrifuge at $200 \times g$ for 5 min, and remove the supernatant.
7. Fix HL-60 cells by intermittent shaking with 75 % ethanol for 30 min at 4 °C.
8. Wash cells twice with PBS and centrifuge $1{,}258 \times g$ for 10 min.
9. Resuspend 1×10^6 cells in 1 mL of PBS.
10. Add 10 μL of Triton X-100 at a final concentration of 1 % and incubate for 30 min at 4 °C to permeabilize cells.
11. Wash twice with PBS and centrifuge at $1{,}258 \times g$ for 10 min.
12. Resuspend cells in 5 % nonfat dry milk-containing PBS and block for 1 h at RT.
13. Add primary antibodies specific to β-actin, TRAF2 and RIP1 at a dilution of 1:100 and incubate overnight at 4 °C (*see* **Note 11**).
14. Wash cells three times with PBS and centrifuge at $1{,}258 \times g$ for 10 min.
15. Add FITC- or TRITC-conjugated corresponding secondary antibodies at a dilution of 1:100 and incubate for 1 h at 37 °C in dark (*see* **Note 11**).
16. Wash cells three times with PBS and centrifuge at $1{,}258 \times g$ for 10 min.
17. Add PI to cells at a final concentration 50 μg/mL and incubate at 4 °C for 5 min to stain nuclei.

18. Wash cells three times with PBS and centrifuge at 1,258 × *g* for 10 min.
19. Resuspend cells in 15 μL of 10 % glycerol PBS.
20. Add 10 μL of 1 % polylysine PBS on a clean slide and dry by air.
21. Add 15 μL of stained cells onto the slide and cover it with a coverslip.
22. Observed under a confocal microscope FU5000.

3.2.2 Actin Polymerization Assay

To observe the actin polymerization, Rhodamine/FITC-phalloidin can be used to stain F-actin. Since one phalloidin specifically bind to one actin monomer of F-actin, it is not only a useful tool for investigating the distribution of F-actin, the amount of fluorescence can also be used as a quantitative measurement of filamentous actin.

1. Harvest HL-60 cells after stimulation (*see* **steps 1–5** in Subheading 3.2.1).
2. Wash cells twice with pre-cold PBS, centrifuge at 200 × *g* for 5 min, and remove the supernatant.
3. Fix cells with 3 % paraformaldehyde by intermittent shaking for 20 min at RT.
4. Wash cells three times with PBS and centrifuge at 1,258 × *g* for 10 min.
5. Permeabilize cells with 0.3 % of Triton X-100 PBS for 3 min at 4 °C.
6. Wash cells twice with PBS.
7. Incubate cells for 40 min at 37 °C with FITC-phalloidin or Rhodamine-phalloidin at a dilution of 1:200 (*see* **Note 12**).
8. Wash cell three times with PBS and centrifuge at 1,258 × *g* for 10 min.
9. Stain nuclei in 50 μg/mL of PI at 4 °C for 5 min.
10. For confocal microscopy, perform **steps 18–22** in Subheading 3.2.1.
11. For quantitation of filamentous actin, resuspend cells in 0.5 mL of PBS and analyze by a FACSCalibur 440E flow cytometer.

3.3 Detection for Signal Transduction

3.3.1 Cytoplasmic, Nuclear, and Total Protein Preparation

1. Preparation of cytoplasmic and nuclear protein.
   (a) Harvest HL-60 cells after stimulation (*see* **steps 1–5** in Subheading 3.2.1).
   (b) Resuspend cells 1×10^6 in 200 μL of pre-cold lysis buffer A and incubate on ice for 10 min by intermittently shaking.
   (c) Add NP-40 at a final concentration of 0.5 % and vortex for 10 s to break down the cytoplasmic membrane.

(d) Centrifugate at 10,000 × *g* for 5 min at 4 °C to separate the cytoplasmic protein from the nucleus.

(e) Collect the supernatant and quick freeze the sample or aliquots on dry ice and store at −80 °C. Set some supernatant aside to determine the protein concentration by a Bradford protein assay. The cytoplasmic protein used to detect IκB and NF-κB p65 by Western blot analysis.

(f) Wash the nuclei three times with ice-cold lysis buffer A and centrifuge 10,000 × *g* for 5 min at 4 °C.

(g) Resuspend the nuclei in 50 μL of ice-cold lysis buffer C and incubated on ice for 30 min by intermittently shaking.

(h) Centrifuge at 12,000 × *g* for 20 min at 4 °C.

(i) Collect the supernatant and quick-freeze the sample or aliquots on dry ice and store at −80 °C besides some supernatant for a Bradford protein assay. The nuclear protein is used for the detection of nuclear translocation of NF-κB p65 by Western blot.

2. Preparation of total protein.

   (a) Harvest cells after stimulation (*see* **steps 1–5** in Subheading 3.2.1).

   (b) Lyze 1 × 10⁶ cells in 0.2 mL of pre-cold total protein lysis buffer on ice for 15 min by intermittently shaking.

   (c) Centrifuge at 13,000 × *g* for 20 min at 4 °C.

   (d) Collect supernatant and quick freeze the sample or aliquots on dry ice and store at −80 °C beside some supernatant for a Bradford protein assay. Total protein is used to detect apoptotic signaling pathway, RIP1, caspase-8, cFLIP, and GAPDH (as a loading control) by Western blot analysis.

3.3.2 Immunoprecipitation (IP)

1. Harvest HL-60 cells after stimulation (*see* **steps 1–5** in Subheading 3.2.1).
2. Lyze cells in 3 mL of pre-cold IP lysis buffer and incubate on ice for 20 min by intermittently shaking.
3. Centrifuge at 13,000 × *g* for 20 min at 4 °C.
4. Collect supernatant, add 1.0 μg of murine IgG and 20 μL of Protein A agarose and incubate at 4 °C for 30 min on an orbital shaker to preclear the cell lysate.
5. Centrifuge at 1,600 × *g* at 4 °C for 10 min.
6. Collect supernatant and determine protein concentration by a Bradford protein assay.
7. Incubate 500 μg of lysate with 6 μg of TNFR2 mAb in 500 μL of pre-cold IP lysis buffer at 4 °C for 2 h (*see* **Note 13**).

8. Add 20 μL of protein A agarose and incubate in rotation at 4 °C overnight.
9. Wash pellet three times with pre-cold IP lysis buffer, centrifuge at 1,600 × g for 10 min at 4 °C.
10. Gently remove the supernatant to avoid aspirating the protein A agarose beads.
11. Add 30 μL of 1× SDS sample buffer and denature at 100 °C for 10 min.
12. Centrifuge at 1,600 × g for 5 min at 4 °C.
13. Collect the supernatant containing immunoprecipitated proteins.
14. Freeze the samples at −20 °C for later use. Frozen samples should be reboiled for 5 min directly prior to loading on a gel.

3.3.3 Western Blot Analysis

1. Add one volume of 4× SDS sample buffer to three volumes of total, nuclear, or cytoplasmic protein solution.
2. Boil at 100 °C for 5 min to denature protein.
3. Load 50 μg of total, nuclear, cytoplasmic, immunoprecipitated protein or prestained molecular weight standards into the wells of a SDS-polyacrylamide gel (*see* **Note 14**).
4. Run the electrophoresis first at 50 V on a 5 % stacking gel until the samples enter 12 % resolving gel and then continue at 80–100 V till the blue front is at the bottom of gel.
5. Pre-wet the following materials mentioned in **step 6** in transfer buffer.
6. Sandwich the proteins-containing gel covered with a precut PVDF membrane between two stacks of Whatman papers (six sheets each stack) without air bubbles.
7. Transfer proteins from the gel onto the PVDF membrane by electroblotting in transfer buffer at 200 mA for 2 h on ice.
8. Block the membranes overnight at 4 °C with 5 % of nonfat dry milk in TBS-T.
9. Probe the membranes for 2 h at RT with different antibodies, including anti-TRAF2, anti-RIP1, anti-caspase-8, anti-cFLIP, anti-TNFR2, anti-IκB, anti-NF-κB p65, anti-β-actin, or anti-GAPDH antibody at a dilution of 1:1,000 (*see* **Note 15**).
10. Wash the membranes three times with TBS-T on an orbital shaker, 5 min each time.
11. Incubate the membranes for 1 h at RT with HRP-conjugated anti-mouse IgG, anti-rabbit IgG, or anti-goat IgG antibody at a dilution of 1:10,000.
12. Repeat **step 10**.

13. Mix an equal volume of ECL detection solution 1 with detection solution 2, then give the mixture onto membrane and wait for 1–5 min.
14. Image bands on the membrane by Kodak image station 4,000 mm.

3.3.4 Analysis of NF-κB Activity by ELISA [21]

1. Harvest HL-60 cells after stimulation (*see* **steps 1–5** in Subheading 3.2.1).
2. Lyse 2×10^6 treated or untreated HL-60 cells in 50 μL of lysis buffer.
3. Incubate HL-60 cells on ice for 10 min by intermittently shaking.
4. Centrifuge at 4 °C for 20 min at 13,000 × *g*, collect the supernatants, and freeze samples at −80 °C.
5. Dissolve two synthesized single-stranded oligonucleotide chains with ddH_2O at a final concentration of 4 μM, respectively.
6. Mix the two chains at a ratio of 1:1, denature at 94 °C for 10 min, and then allow them to anneal at RT to form the double-stranded probe.
7. Dilute the double-stranded probe with PBS at a final concentration of 2 pM.
8. Add 50 μL of double-stranded probe to streptavidin-coated 96-well plates and incubate at 37 °C for 1 h, allowing the probe to bind streptavidin by its conjugated biotin.
9. Wash the plates three times with PBS-T, 5 min each time.
10. Mix 20 μL of whole-cell lysate containing 5 μg of protein with 30 μL of binding buffer and add the mixture to well.
11. Incubate for 1 h at RT on an orbital shaker.
12. Repeat **step 9** in this section.
13. Add 100 μL of NF-κB p65 mAb at a dilution of 1:1,000 to each well and incubate for 1 h at RT.
14. Repeat **step 9** in this section.
15. Add 100 μL of peroxidase-conjugated anti-mouse IgG at a dilution of 1:1,000 to each well and incubate for 1 h at RT.
16. Repeat **step 9** in this section.
17. Dilute 50 μL of 0.2 % TMB with 1 mL of substrate buffer and 2.5 μL of 30 % H_2O_2.
18. Add 100 μL of substrate solution to each well and incubate for 5–15 min at RT in dark (*see* **Note 16**).
19. Add 100 μL of stop solution to each well to stop color reaction.
20. Read OD values at 450 nm on an ELISA reader using a reference wavelength of 655 nm.

4 Notes

1. For tmTNF-α-mediated cytotoxicity, a pair of effector cell and target cell can also be suspension cells or adherent cells, because effector cells should be fixed prior to coculture with target cells, and later detection of cell viability by MTT is only for the living target cells rather than fixed effector cells. However, it is difficult to separate from each other for many other tests when both effector and target cells are suspension or adherent.
2. As tmTNF-α can be processed to sTNF-α, noncleavable tmTNF-α mutant (deletion of TACE cleavage site) is often used for stable transfection of cells. However, we found that this mutant functions weaker than wtTNF-α. To exclude changes in signal transduction of the mutant, we chose wtTNF-α-transfected NIH3T3 cells as effector cells. Furthermore, some tumor cell lines (such as Raji, a B lymphoma cell line, and MDA-MD-231, a breast cancer cell line) that express tmTNF-α at high levels [22, 23] can also be used as the source of tmTNF-α.
3. CytD is dissolved with DMSO to 2 mM CytD as a stock solution, stored at −20 °C. It should be fresh diluted at a final concentration of 2 μM CytD in medium at the time of use.
4. These lysis buffers can be stored at 4 °C for up to 6 months or −20 °C for up to 1 year. PMSF, aprotinin, and leupeptin should be added fresh at the time of use.
5. 30 % Acrylamide/bis is filtered through a 0.45 μm filter and stored at 4 °C in a bottle wrapped with aluminum foil. 10 % Ammonium persulfate should be prepared fresh at the time of use.
6. During fixation, shake tube or plate intermittently to completely fix cells, because fixed cells are easy to form agglomerate. Further, as the ratio of effector and target cells is 10:1, the wells of the 96-well microtiter plate are too small to allow such an amount of NIH3T3 cells to adhere. Therefore, NIH3T3 cells are fixed in suspension without adherence.
7. Usually, cells are centrifuged at 200 × g for 5 min. The weight of fixed cells becomes much lighter so that the speed and time of centrifugation should be increased at 1,258 × g for 10 min.
8. As fixed effector cells cannot reduce MTT into colored formazan product by mitochondrial succinate dehydrogenase, which occurs only in metabolically active cells, the detection of cell viability by MTT is only for the living target cells.
9. Wash the fixed NIH3T3 cells in the plate with PBS by gently shaking the plate, instead of pipetting. Then, remove PBS carefully from each well without disturbing the adherent effector cells on the bottom of the wells.

10. We corroborated our key data (tmTNF-α-mediated apoptosis) obtained with CytD using an alternative pharmacological agent LatA. Although both agents are F-actin-destabilizing compounds, they act a little differently. CytD caps the barbed end of actin filaments and severs them, while LatA sequesters actin monomers, inhibiting actin polymerization.
11. It is very important that the host species of primary antibodies should be different from each other, and the different fluorescence-conjugated secondary antibodies should be matched with the primary antibodies.
12. There are several fluorophores that are conjugated with phalloidin, such as Alexa Fluor 350, Alexa Fluor 488, Rhodamine, Alexa Fluor 594, and Alexa Fluor 647. The corresponding excitation and emission filters should be chosen to distinguish the fluorophores from the different fluorescence-conjugated antibody-stained proteins. If PI or DAPI staining is followed, Alexa Fluor 350-phalloidin cannot be used.
13. As TNFR2 has 28 % homology with TNFR1 mostly in their extracellular domain with four randomly repeated cysteine rich motifs, one should make sure that the antibody specific to TNFR2 is not cross-reactive to TNFR1 to avoid immunoprecipitating signal molecules with TNFR1.
14. In addition to the loading dose of protein, the volume of samples loaded in each lane should also be the same.
15. Because CytD has been reported to induce β-actin expression in a variety of cell types [24], GAPDH can be used to replace β-actin as a loading control.
16. The substrate solution should be prepared immediately prior to use. Substrate is light sensitive so that the substrate solution should be in a dark bottle or in a vessel wrapped with aluminum foil.

References

1. Black RA, Rauch CT, Kozlosky CJ, Peschon JJ, Slack JL, Wolfson MF et al (1997) A metalloproteinase disintegrin that releases tumor-necrosis factor-alpha from cells. Nature 385:729–733
2. Grell M, Douni E, Wajant H, Löhden M, Clauss M, Maxeiner B et al (1995) The transmembrane form of tumor necrosis factor is the prime activating ligand of the 80 kDa tumor necrosis factor receptor. Cell 83:793–802
3. Cabal-Hierro L, Lazo PS (2012) Signal transduction by tumor necrosis factor receptors. Cell Signal 24:1297–1305
4. Peck R, Brockhaus M, Frey JR (1989) Cell surface tumor necrosis factor (TNF) accounts for monocyte- and lymphocyte-mediated killing of TNF-resistant target cells. Cell Immunol 122:1–10
5. Shi W, Li Z, Gong F et al (1998) Comparison of the cytocidal effect induced by transmembrane and secreted TNF-α. Chin J Microbiol Immunol 18:499–504
6. Zhang S, Liu T, Liang H, Zhang H, Yan D, Wang N et al (2009) Lipid rafts uncouple surface expression of transmembrane TNF-alpha from its cytotoxicity associated with ICAM-1 clustering in Raji cells. Mol Immunol 46: 1551–1560
7. Bunnell SC, Kapoor V, Trible RP, Zhang W, Samelson LE (2001) Dynamic actin polymer-

ization drives T cell receptor-induced spreading: a role for the signal transduction adaptor LAT. Immunity 14:315–329

8. Hao S, August A (2005) Actin depolymerization transduces the strength of B-cell receptor stimulation. Mol Biol Cell 16: 2275–2284
9. Ingram AJ, James L, Cai L, Thai K, Ly H, Scholey JW (2000) NO inhibits stretch-induced MAPK activity by cytoskeletal disruption. J Biol Chem 275:40301–40306
10. Kawamura S, Miyamoto S, Brown JH (2003) Initiation and transduction of stretch-induced RhoA and Racl activation through caveolae: cytoskeletal regulation of ERK translocation. J Biol Chem 278:31111–31117
11. Kustermans G, El Benna J, Piette J, Legrand-Poels S (2005) Perturbation of actin dynamics induces NF-kappaB activation in myelomonocytic cells through an NADPH oxidase-dependent pathway. Biochem J 387: 531–540
12. Are AF, Galkin VE, Pospelova TV, Pinaev GP (2000) The p65/RelA subunit of NF-kappa B interacts with actin-containing structure. Exp Cell Res 256:533–544
13. Fazal F, Minhajuddin M, Bijli KM, McGrath JL, Rahman A (2007) Evidence for actin cytoskeleton-dependent and -independent pathways for RelA/p65 nuclear translocation in endothelial cells. J Biol Chem 282: 3940–3950
14. Prendergast GC (2001) Actin' up: RhoB in cancer and apoptosis. Nat Rev Cancer 1:162–168
15. Shive MS, Brodbeck WG, Colton E, Anderson JM (2002) Shear stress and material surface effects on adherent human monocyte apoptosis. J Biomed Mater Res 60:148–158
16. Jaffe AB, Hall A (2005) Rho GTPases: biochemistry and biology. Annu Rev Cell Dev Biol 21:247–269
17. Charest PG, Firtel RA (2007) Big roles for small GTPases in the control of directed cell movement. Biochem J 401:377–390
18. Chen H, Xiao L, Zhang H, Liu N, Liu T, Liu L et al. (2011) The involvement of β-actin in the signaling of transmembrane TNF-α-mediated cytotoxicity. J Leukoc Biol 89:917–926
19. Li Q, Li L, Li Z, Gong F, Feng W, Jiang X et al (2002) Antitumor effects of the fibroblasts transfected TNF-α gene and its mutants. J Huazhong Univ Sci Technolog Med Sci 22:92–95
20. Mazur L, Augustynek A, Bochenek M (2002) Flow cytometric estimation of the plasma membrane diversity of bone marrow cells in mice treated with WR-2721 and cyclophosphamide. Toxicology 171:63–72
21. Jin S, Lu D, Ye S, Ye H, Zhu L, Feng Z et al (2005) A simplified probe preparation for ELISA-based NF-κB activity assay. J Biochem Biophys Methods 65:20–29
22. Zhang H, Yan D, Shi X, Liang H, Pang Y, Qin N et al (2008) Transmembrane TNF-α mediates "forward" and "reverse" signaling, inducing cell death or survival via the NF-κB pathway in Raji Burkitt lymphoma cells. J Leukoc Biol 84:789–797
23. Yan D, Qin N, Zhang H, Liu T, Yu M, Jiang X et al (2009) Expression of TNF-α leader sequence renders MCF-7 tumor cells resistant to the cytotoxicity of soluble TNF-α. Breast Cancer Res Treat 116:91–102
24. Sympson CJ, Geoghegan TE (1990) Actin gene expression in murine erythroleukemia cells treated with cytochalasin D. Exp Cell Res 189:28–32

Chapter 7

Investigating the Protective Role of Death Receptor 3 (DR3) in Renal Injury Using an Organ Culture Model

Rafia S. Al-Lamki

Abstract

Death receptor 3 (DR3; also designated as Wsl-1, Apo3, LARD, TRAMP, TNFRSF25, and TR3) is a member of the tumor necrosis factor (TNF) receptor superfamily that has emerged as a major regulator of inflammation and autoimmune diseases. DR3 contains a homologous intracellular region called the death domain (DD) that can bind adaptor proteins, which also contain a DD, initiating cellular responses such as caspase activation and apoptotic cell death. However, in other circumstances DR3 can initiate induction of transcription genes and gene products that can prevent cell death from occurring. Our laboratory has reported an inducible expression of DR3 in human and mouse tubular epithelial cells in renal injury, but its function in these setting still remains unclear. To directly manipulate and evaluate the role of DR3 in vivo, I have used an in vitro organ culture (OC) model, which I have developed in our laboratory. In this chapter, I will describe in detail the OC model used to study the role of DR3 in renal injury and discuss its advantages and limitations. In my hands, the OC model has proven to be an efficient tool for studying human cell heterogeneity, basal and regulated receptor expression, signalling pathways, and various biological responses not readily achievable in traditional cell culture models. Various assays can be carried out on organ cultures including histology, biochemistry, cell biology, and molecular biology, which will not be described in detail in this chapter.

Key words Death receptor 3 (DR3), TL1A, TNF, Organ culture, Renal injury, Human, Mouse

1 Introduction

Tissues were first cultured before the turn of the century and since that time a multitude of techniques, each designed for the solution of a particular problem, has been devised [1–12]. All of these sources with different organs from a variety of animals exhibit the diversity of the organ culture model. New models are required to investigate cellular cross-talk between different cell types and to construct complex in vitro models to properly study tissue, organ, and system interaction without resorting to animal experiments. I have developed an in vitro organ culture (OC) model in our laboratory, which has been, and continues to be, very valuable in the study of in situ functions of specific molecules by experimental

Jagadeesh Bayry (ed.), *The TNF Superfamily: Methods and Protocols*, Methods in Molecular Biology, vol. 1155,
DOI 10.1007/978-1-4939-0669-7_7, © Springer Science+Business Media New York 2014

addition of exogenous cytokines, growth factors, and pharmacological intervention [7, 8, 13]. Physiological conditions and exposures can be carefully controlled and manipulated to test hypotheses and explore biochemical and molecular mechanisms of action. Organ cultures can be subsequently used for various assays including histology, biochemistry, cell biology, and molecular biology.

Death receptor 3 (DR3) was initially reported to be restricted to lymphoid cells [14–16]; however, our laboratory has reported its preferential expression in tubular epithelial cells (TECs) in human renal biopsy specimens with evidence of acute allograft rejection and ischemic injury [17]. To further evaluate the role of DR3 in renal injury, we have used tissue from wild type ($DR3^{+/+}$) and congenitally deficient DR3 ($DR3^{-/-}$) mice maintained in organ culture and treated the cultures with cisplatin (*cis*-diamminedichloroplatinum II; CDDP) to induce controlled nephrotoxicity as a model of renal injury [18, 19]. Addition of cisplatin to cultures of $DR3^{+/+}$ resulted in the induction of DR3, its principal ligand TL1A (also known as TNFSF15), TNF, apoptotic cell death, and NF-κB activation in TECs. In comparison, cisplatin induced an increased apoptotic cell death and less pronounced activation of NFκB in TECs in cultures from $DR3^{-/-}$ mice. We have also utilized the OC model to further evaluate the responses of human and mouse kidney tissue to TL1A and to determine the importance of DR3 in response to TL1A, using TNF as control. Addition of TL1A to cultures of human kidney and $DR3^{+/+}$ mice induced caspase-3 activity, albeit less compared to TNF-treatment cultures whereas the activation of NFκB was markedly enhanced. In contrast, TL1A did not elicit activation of NF-κB or activation of cleaved caspase-3 in organ cultures from $DR3^{-/-}$ mice but treatment with TNF did [17]. Concomitant activation of caspase-3 and NF-κB is typical hallmark of death receptor signals, especially those induced by TNF receptor superfamily members [20]. Collectively these data indicate that signalling through DR3 is protective rather than pro-death in injured renal TECs and that DR3 signalling mitigates cisplatin-induced nephrotoxicity possibly by antagonizing pro-apoptotic signals induced by TNF. This is because DR3 is a partial agonist compared to TNF. The OC model can also be manipulated to intervene in signalling pathways using specific inhibitors. For example in our previous studies we have demonstrated TL1A induction of TNFR2 expression in TECs in cultures of human and mouse kidney tissue [17]. To further assess if TL1A-induced TNFR2 expression in TECs is dependent on NFκB activation, we pre-treated organ cultures with a NFκB-specific inhibitors (pyrrolidine dithiocarbamate; PDTC [21] or thalidomide [22]) for 30 min before treatment with TL1A or TNF. This effectively inhibited TL1A-mediated activation of NF-κB but did not alter TNFR2 induction on TECs [17], suggesting that TL1A-induction of TNFR2 is independent of NF-κB activation. The results described here have been presented in peer-reviewed papers [17, 23] that form the basis for this chapter and some of the information here has been made available previously. Collectively our OC model has proved to be a useful and versatile tool that can be widely employed in research to study cellular responses in an in vivo-like setting. This model has the potential to significantly reduce animal use by allowing control and manipulation of tissue in vitro. The protocol described in this chapter describes the use of the OC model in investigating the effect of cytokines (TL1A and TNF) and of specific inhibitors in the expression and function of DR3 in kidney tissue. Despite the unprecedented efficiency of the OC model, the results must be interpreted cautiously and validated using proper controls.

2 Materials

2.1 Collection and Preparation of Tissue for Organ Culture (See Note 1)

1. Human Tissue: All experiments using human tissue must be performed with informed consent and approval by the local ethical committee and tissue bank. Tissue for organ culture can be obtained from the uninvolved pole of kidney excised because of renal tumor or from biopsies taken immediately after reperfusion of renal transplants (*see* **Note 2**). The nephrectomy specimens received in histopathology usually are pending diagnosis so it is important to involve a specialist pathologist as a collaborator to assist in sample selection and diagnosis. The pathologist will dissect the nephrectomy specimen longitudinally, exposing the two halves of a kidney and provide you with some tissue from the cortex and from uninvolved part of the tumor (pathologically normal kidney). The pathologist should be cooperative in order to assure that freshness of the tissue is an absolute priority. The need to use extremely fresh tissue, preferably without touching and distortion is of utmost importance. The pathologist, in most part, will offer a small sample of tissue (~3–5 cm^3) at any given time for purpose of research and will keep the rest for diagnostic studies. The fresh kidney tissue is immediately transferred to a sterile 50-mL universal tube, containing ice-cold tissue culture medium (contents are described in Subheading 2) and immediately transported to the laboratory. The time between excision of the specimen and setting up of OC experiments is critical and should not exceed 20 min.
2. Mouse Tissue: All protocols involving animals must be subjected to the local ethical committee for approval and conducted according to the personal, project, and institutional licenses UK Animals (Scientific Procedures) Act 1986. Breed wild-type ($DR3^{+/+}$), heterozygous ($DR3^{+/-}$), and knockout ($DR3^{-/-}$)

mice used in the OC experiments from heterozygous parents. Cross once DR3$^{-/-}$ mice [24] into a CD1 background, and cross the F1 heterozygote progeny to yield DR3$^{+/+}$ and DR3$^{-/-}$ littermates. Breed DR3$^{-/-}$ mice into a CD1 background over >10 generations. Euthanize the 8–13 weeks mice by CO_2 inhalation, collect the kidneys, and immerse immediately in ice-cold tissue culture medium and transport to the laboratory (*see* **Note 3**).

2.2 Tissue Culture Medium, Solutions, and Equipments for In Vitro Organ Culture

1. Tissue Culture Medium: M199 Medium, 10 % heat-inactivated fetal calf serum, 100 units/mL penicillin, 100 μg/mL streptomycin, 2 mM L-glutamine. This medium can be stored at 4 °C for up to 6 months.
2. Sterile flat-bottomed 96-well tissue culture plates.
3. Human or mouse recombinant TL1A (R&D Systems, UK).
4. Human or mouse recombinant TNF-α (R&D Systems).
5. Pyrrolidine dithiocarbamate (PDTC, Merck Biosciences Ltd, UK).
6. Thalidomide (Sigma-Aldrich, UK).
7. Sterile plastic disposable pipettes or autoclaved fine metal forceps.
8. Sterile polystyrene petri dishes 90 × 15 mm.
9. Sterile Dulbecco's phosphate-buffered saline (PBS): PBS (137 mM NaCl, 2.7 mM KCl, 8 mM Na_2HPO_4, 2 mM KH_2PO_4) with 0.5 mM $MgCl_2$, 0.9 mM $CaCl_2$.
10. Multi-channel pipette or single pipettes and pipette tips (1–1,000 μL).
11. Laminar flow Biological Safety Cabinet.
12. 5 % CO_2 incubator maintained at 37 °C.
13. 4 °C fridge and −20 °C freezer for storage of the reagents and of cultures following OC.
14. Water bath.

3 Methods

3.1 Dissection of Tissue for Organ Culture

1. Upon arrival in the laboratory, decant the 3–5 cm^3 piece of fresh human kidney tissue in a sterile petri dish filled with tissue culture medium. Tissue can be rinsed in sterile PBS prior to dissection to remove blood.
2. Using a sterile forceps and a sharp scalpel blade, gently dissect the human kidney tissue into small pieces of less than 1 mm^3 in thickness to ensure optimum diffusion of reagents. For a whole mouse kidney, grasp the kidney using a sterile forceps and

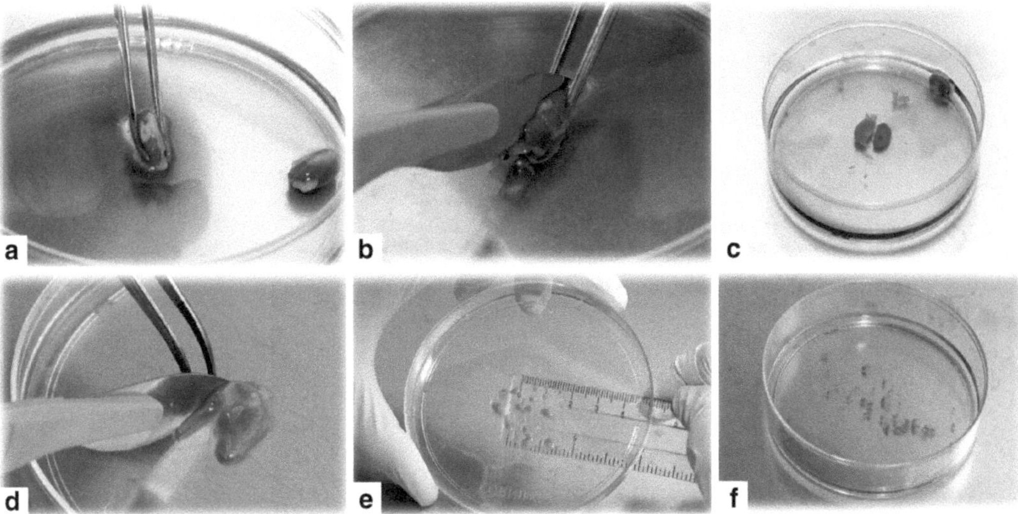

Fig. 1 Preparation of tissue for organ cultures. (**a**, **b**) Mouse kidney is held firmly using sterile forceps and sliced vertically into two halves exposing the cortex and medulla (**c**). Each half is further dissected into small fragments ≤1 mm^3 and immersed in tissue culture medium (**d–f**)

dissect it longitudinally into two halve, then into further pieces of less than 1 mm^3 as depicted in Fig. 1a–f (*see* **Note 4**). Do not put any pressure on the tissue while cutting, and make sawing movements with the razor blade. Keep tissue slices covered with the tissue culture medium and cut multiple 1 mm × 1 mm strips under fluid (*see* **Note 5**). One should avoid subjecting the tissue to dryness or dry conditions; keep it wet and well immersed in fluid. A dissecting microscope is useful in selecting areas of interest if available in your laboratory.

3.2 Setting Up Organ Culture

1. In a sterile flat-bottomed 96-well plate dispense 200 μL/well of tissue culture medium in duplicate ensuring complete coverage of the each well bottom as depicted in Fig. 2a (*see* **Notes 6** and 7). The medium in the 96-well plate(s) can be pre-warmed in 37 °C and sealed with cling film to avoid evaporation of the medium. It is important to have sufficient medium in each well to enhance diffusion of gases and ensure there is no physical compression effect of surface tension.
2. Immediately immerse the dissected 1 mm^3 pieces of human or mouse kidney tissue in each well containing the 200 μL of medium in the 96-well plate (do not place more than four small tissue pieces in each well to ensure optimal diffusion of reagents) and allow them to sink to the bottom as depicted in Fig. 2b–d (*see* **Notes 6** and 7). Cultures must be run in duplicate or triplicate to allow group comparisons and to assess the reliability and reproducibility of assays (*see* **Note 8**).

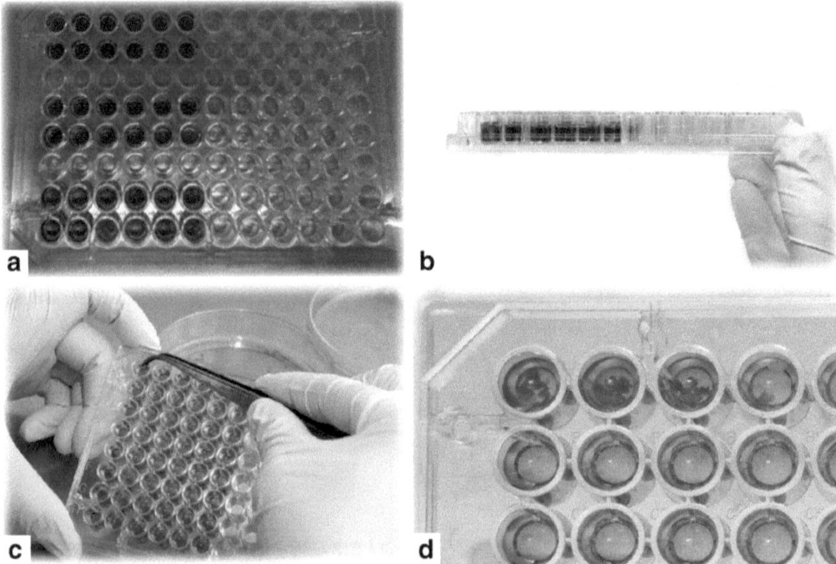

Fig. 2 Setting up organ culture. Dispense 200 μL per well of the tissue culture medium in a sterile flat-bottomed 96-well plate ensuring complete coverage of each well bottom (**a**, **b**). Within each well immerse approximately two to four pieces of tissue fragments, ensuring the pieces are well immersed in fluid (**c**, **d**)

3. To analyze the effects of TL1A and TNF on DR3 in cultures of human or mouse kidney tissue, either leave the cultures in medium alone (untreated controls) or treat with species-specific recombinant TL1A or -TNF as depicted in Fig. 3a, b. If testing for the first time it is advisable to set up a dose–titration curve of the reagents of interest. In this instance, the optimal response of human and mouse kidney tissue to recombinant TL1A or TNF was worked out in our laboratory using dose and time titrations, with maximum response achieved at a concentration of 0.2 μg/mL for TL1A and 10 ng/mL for TNF. Prepare solutions only as much as is required immediately before use. Cover the plate with cling film and incubated immediately in 5 % CO_2 for 3 h at 37 °C (24 h is the maximum incubation period for tissue cultures after which rapid deterioration of tissue is observed).
4. To study the effects of cisplatin-induced renal injury in mouse kidney, place the pieces of fresh tissue from $DR3^{+/+}$ or $DR3^{-/-}$ mice (not more than 1 mm^3, in duplicate) in flat-bottomed 96-well plates containing 200 μL of the tissue culture medium. The tissue is either left in medium alone (untreated controls) or immersed in wells with medium containing 80 μmol/L of cisplatin and incubated in 5 % CO_2 for 3 h at 37 °C as illustrated in Fig. 3c. (A dose and time-response curve should be carried out to work out the optimal response of cisplatin).

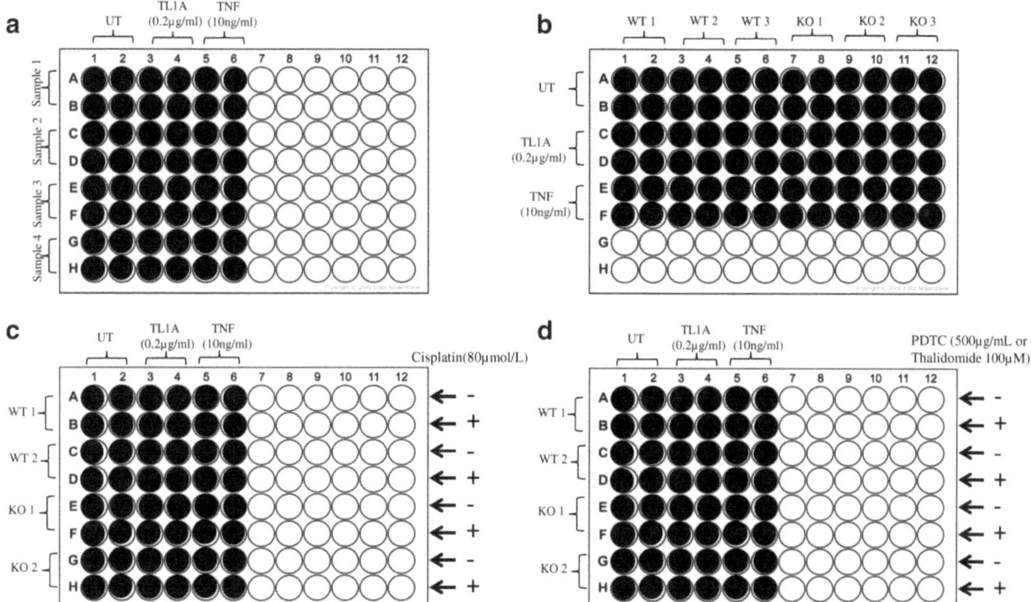

Fig. 3 Analyzing the effects of treatments in cultures of human or mouse kidney tissue. (**a**, **b**) To analyze the effects of TL1A and TNF on DR3 in human kidney tissue (samples 1–4) or DR3$^{+/+}$ (WT) or DR3$^{-/-}$ mice (KO) mice kidney tissue, samples (run in duplicate or triplicate) are immersed in 96-well plates containing 200 μL per well and either left in medium alone (untreated; UT) or treated with species-specific recombinant 0.2 μg/mL TL1A or 10 ng/mL TNF (A dose and time-response curve should be carried out to work out the optimal response of TL1A and TNF). Plates are then covered with a cling film and incubated for 3 h at 37 °C. (**c**) To analyze the effects of cisplatin-induced renal injury, pieces of kidney from WT or DR3 KO mice are either left in medium alone (UT) or treated with 80 μmol/L cisplatin and incubated at 37 °C for 3 h. (A dose and time-response curve should be carried out to work out the optimal response of cisplatin). (**d**) To determine the downstream signalling molecules involved in TL1A or TNF ligation, cultures of kidney tissue from WT or DR3 KO mice can be treated with 50 μg/mL PDTC or 100 μM thalidomide (specific inhibitors of NFκB) for 30 min prior to treatment with TL1A or TNF and plate incubated for 3 h at 37 °C

5. To delineate the downstream signalling molecules involved in TL1A or TNF ligation, treat the organ cultures of kidney tissue from human or DR3$^{+/+}$ or DR3$^{-/-}$ mice with 50 μg/mL of PDTC or 100 μM of thalidomide (specific inhibitors of NFκB) for 30 min prior to treatment with TL1A or TNF as depicted in Fig. 3d. Incubate all cultures in 5 % CO_2 for 3 h at 37 °C.
6. At the end of the 3 h incubation, remove all the cultures and process immediately and according to subsequent experiments. Various assays can be carried out on organ cultures including histology, biochemistry, cell biology, and molecular biology (*see* **Note 9**). Cultures can be fixed in paraformaldehyde and glutaraldehyde for light microscopy and electron microscopy studies and snap-frozen in liquid nitrogen for cell and molecular biology analysis as depicted in Fig. 4.

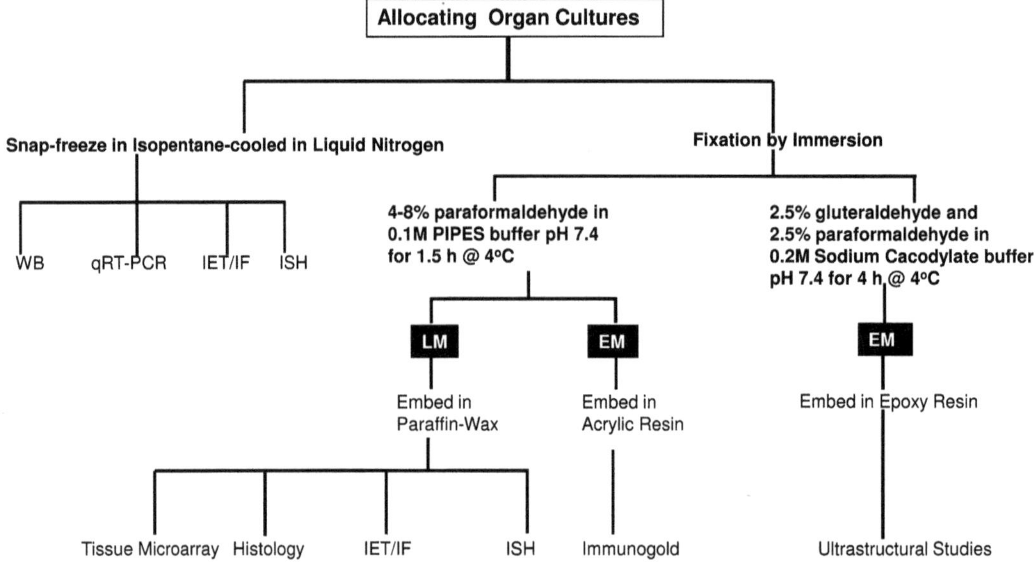

Fig. 4 A schematic diagram shows the investigations that can be carried out on organ cultures. Cultures can either be snap-frozen in isopentane-cooled liquid nitrogen for cell and molecular biology analysis or immediately fixed by immersion in paraformaldehyde and/or glutaraldehyde for light microscopy (LM) and electron microscopy (EM) studies. The strength of the fixative used will depend on the subsequent investigation(s). (*IET* immunoenzyme technique, *qRT-PCR* quantitative real-time polymerase chain reaction, *IF* immunofluorescence, *ISH* in situ hybridization, *WB* western blotting)

4 Notes

1. The development of OC system offers considerable advantages to traditional cell culture methods as it provides an invaluable in situ bioassay. It can be used to study cell heterogeneity and cell interactions, basal and regulated receptor expression, signalling pathways, and various biological responses not readily achievable in traditional cell culture models. Genetic manipulation, restricted primarily to rodents, can be directly applied to an all-human system. A technical difficulty of the OC model is the speed at which the pieces of tissue must be in culture. Samples have to be cultured as rapidly as possible (preferably less than 20 min from cessation of blood supply to culture) to preserve maximum viability.
2. Through the control of temperature and CO_2, which can be fine-tuned for a particular need, the OC model mimics the natural environment inside the body, offering direct experimentation on human tissues. In situ geometries are well preserved, which allows investigation of basic biological questions of human cell–cell and cell–microenvironment interactions that not easily obtainable using isolated cells in culture.

The OC model also addresses how conditions affect general organ development, differentiation, and maintenance of the differentiated state.

3. By removing the kidney from the rest of the body, systemic effects from other areas of the body are eliminated. Removing the systemic effects gives us a better understanding of what is actually occurring in the target tissue through the injury process. The removal of systemic effects is not possible in an in vivo model, although one can try to mimic these changes to test, for example, which of these effects has the most influence on the process of injury.
4. Embedding in native matrix allows cells to attain a physical conformation that maximizes secretion of biochemical regulators and minimizes immunogenicity.
5. The availability of only small amounts of tissue for analysis is the major limitation of the OC model. The size of the cultured samples, which must be less then 1 mm^3 and preferably uniform. Although 1 mm^3 tissue in organ culture yields a large number of cells of interest in a compact setting compared to the amounts of cells that need to be cultured to achieve the same numbers, it is limited to a selective area. Only a selective area of the kidney can be cultured at one time and therefore the inexperienced scientists are advised to familiarize themselves with the tissue of study before embarking on the OC model described in this chapter.
6. Using an OC modeling approach, kidney maintained in culture may be subjected to stimulation and the effect of the stimulation assessed in a physiologic milieu maintaining appropriate cell types and numbers within their 3D, communication-intact environment. Cells in OC are exposed to many endogenous signals, involving both humoral and contact-dependent mechanisms, which are *per force* ablated by isolating cells in cultures. The OC model allows study of various cell types that are not readily represented in isolated cell cultures. The presence of a heterogeneous population of cells within a tissue of study and the matrix composition are essential for protein expression and function, and therefore lead to culture results not readily reflecting those obtained from isolated cells in culture.
7. The OC model is very economical, utilizing only a small volume of medium (200 μL per well in a 96-well plate) and minimal reagents per experiment, while maximizing the number of cells available for study.
8. The lack of a vascular system is one reason why the viability of the tissue drops in culture. Culture medium is present, but it cannot replicate a vascular system that delivers minerals and nutrients and removed waste materials that normally occurs in

the body. Additionally, the biochemical environment may not be as nourishing to living cells as a healthy in vivo environment, which contains circulating, cells or factors from other tissues.

9. In my experience, results obtained from in vitro studies of isolated cell cultures and OC do not always correlate, possibly as a reflection of heterogeneity with the OC model whereby the induction of protein expression may be influenced either by neighboring cells in contact or from, some diffusible factors from elsewhere and therefore creates redundancy in vitro but not OC system.

References

1. Alm R, Edvinsson L, Malmsjo M (2002) Organ culture: a new model for vascular endothelium dysfunction. BMC Cardiovasc Disord 2:8
2. Anderson G, Jenkinson EJ (2007) Fetal thymus organ culture. CSH Protoc 2007:pdb prot4808
3. Barr J, Weir AJ, Brendel K, Sipes IG (1991) Liver slices in dynamic organ culture. II. An in vitro cellular technique for the study of integrated drug metabolism using human tissue. Xenobiotica 21:341–350
4. Hart A, Mattheyse FJ, Balinsky JB (1983) An organ culture of postnatal rat liver slices. In Vitro 19:841–852
5. Lambrot R, Coffigny H, Pairault C, Donnadieu AC, Frydman R, Habert R et al (2006) Use of organ culture to study the human fetal testis development: effect of retinoic acid. J Clin Endocrinol Metab 91:2696–2703
6. Martins-Green M, Li QJ, Yao M (2005) A new generation organ culture arising from cross-talk between multiple primary human cell types. FASEB J 19:222–224
7. Messadi DV, Pober JS, Fiers W, Gimbrone MA Jr, Murphy GF (1987) Induction of an activation antigen on postcapillary venular endothelium in human skin organ culture. J Immunol 139:1557–1562
8. Messadi DV, Pober JS, Murphy GF (1988) Effects of recombinant gamma-interferon on HLA-DR and DQ expression by skin cells in short-term organ culture. Lab Invest 58:61–67
9. Schurch W, McDowell EM, Trump BF (1978) Long-term organ culture of human uterine endocervix. Cancer Res 38:3723–3733
10. Sebinger DD, Unbekandt M, Ganeva VV, Ofenbauer A, Werner C, Davies JA (2010) A novel, low-volume method for organ culture of embryonic kidneys that allows development of cortico-medullary anatomical organization. PLoS One 5:e10550
11. Varani J, Perone P, Spahlinger DM et al (2007) Human skin in organ culture and human skin cells (keratinocytes and fibroblasts) in monolayer culture for assessment of chemically induced skin damage. Toxicol Pathol 35:693–701
12. Voisard R, Baur R, Herter T, Krüger U, Breul J, Keller L et al (2007) A perfused renal human organ culture model: impact of monocyte attack. Med Sci Monit 13:CR82–CR88
13. Al-Lamki RS, Wang J, Vandenabeele P, Bradley JA, Thiru S, Luo D et al (2005) TNFR1- and TNFR2-mediated signaling pathways in human kidney are cell type-specific and differentially contribute to renal injury. FASEB J 19: 1637–1645
14. Bodmer JL, Burns K, Schneider P, Hofmann K, Steiner V, Thome M et al (1997) TRAMP, a novel apoptosis-mediating receptor with sequence homology to tumor necrosis factor receptor 1 and Fas(Apo-1/CD95). Immunity 6:79–88
15. Chinnaiyan AM, O'Rourke K, Yu GL, Lyons RH, Garg M, Duan DR et al (1996) Signal transduction by DR3, a death domain-containing receptor related to TNFR-1 and CD95. Science 274:990–992
16. Kitson J, Raven T, Jiang YP, Goeddel DV, Giles KM, Pun KT et al (1996) A death-domain-containing receptor that mediates apoptosis. Nature 384:372–375
17. Al-Lamki RS, Wang J, Tolkovsky AM, Bradley JA, Griffin JL, Thiru S et al (2008) TL1A both promotes and protects from renal inflammation and injury. J Am Soc Nephrol 19: 953–960
18. Pabla N, Dong Z (2008) Cisplatin nephrotoxicity: mechanisms and renoprotective strategies. Kidney Int 73:994–1007
19. Yao X, Panichpisal K, Kurtzman N, Nugent K (2007) Cisplatin nephrotoxicity: a review. Am J Med Sci 334:115–124

20. Ramakrishnan P, Baltimore D (2011) Sam68 is required for both NF-kappaB activation and apoptosis signaling by the TNF receptor. Mol Cell 43:167–179
21. Zeng M, Yan H, Chen Y, Zhao HJ, Lv Y, Liu C et al (2012) Suppression of NF-kB reduces myocardial no-reflow. PLoS One 7:e47306
22. Majumdar S, Lamothe B, Aggarwal BB (2002) Thalidomide suppresses NF-kappa B activation induced by TNF and H_2O_2, but not that activated by ceramide, lipopolysaccharides, or phorbol ester. J Immunol 168:2644–2651
23. Al-Lamki RS, Lu W, Finlay S, Twohig JP, Wang EC, Tolkovsky AM et al (2012) DR3 signaling protects against cisplatin nephrotoxicity mediated by tumor necrosis factor. Am J Pathol 180:1454–1464
24. Wang EC, Thern A, Denzel A, Kitson J, Farrow SN, Owen MJ (2001) DR3 regulates negative selection during thymocyte development. Mol Cell Biol 21:3451–3461

Chapter 8

Analysis of TNF-alpha-Mediated Cerebral Pericyte Remodeling

Jennifer V. Welser-Alves, Amin Boroujerdi, Ulrich Tigges, and Richard Milner

Abstract

As well as being a central regulator of inflammatory and immune-mediated events, TNF-α also influences vascular remodeling, resulting in alterations in the structure and function of blood vessels. In addition to endothelial cells, pericytes are another type of vascular cell that significantly contribute to the development, maturation, stabilization, and remodeling of blood vessels. To investigate the regulatory influence of different factors on pericyte behavior, we recently described a novel yet simple approach of isolating and culturing highly pure, high density cultures of mouse brain pericytes. In this chapter, we briefly describe this culture system, as well as methods for examining different aspects of pericyte behavior, including cell adhesion, cell migration, and cell proliferation. These assays can be used to examine the influence of TNF-α or any other factor on pericyte behavior.

Key words TNF-α, Pericyte, Brain endothelial cell (BEC), Culture, Purity, Adhesion, Migration, Proliferation

1 Introduction

TNF-α is a pleiotropic cytokine that influences a wide variety of physiological and pathological processes. In addition to being a central regulator of inflammatory and immune-mediated events, TNF-α also has a strong regulatory influence on vascular remodeling, which culminates in alterations in the structure and function of blood vessels. For instance, TNF-α promotes angiogenic sprouting in the rat cornea and chick chorioallantoic membrane [1]. In a mouse model of airway inflammation, TNF-α levels and endothelial expression of TNF receptor 1 (TNF-R1) were increased and inhibition of this pathway blocked vascular remodeling [2]. Furthermore, at the cellular level, TNF-α promotes endothelial cell proliferation [3], migration and tube formation [1].

In addition to endothelial cells, pericytes are another type of vascular cell that significantly contribute to the development, maturation, stabilization, and remodeling of blood vessels [4].

Jagadeesh Bayry (ed.), *The TNF Superfamily: Methods and Protocols*, Methods in Molecular Biology, vol. 1155, DOI 10.1007/978-1-4939-0669-7_8, © Springer Science+Business Media New York 2014

In the central nervous system (CNS) pericytes participate in the regulation of the blood–brain barrier (BBB) and process of angiogenesis [5–7]. Pericyte dysfunction has also been shown to be linked to several distinct vascular pathologies, including hypertension [8, 9], diabetic microangiopathy, and other diseases in which disordered angiogenic remodeling is a major part [4, 10]. Therefore it is important to improve our understanding of pericyte biology with the goal of developing novel therapeutics for the treatment of different diseases with a strong vascular component.

Up until recently pericytes were difficult to study because of the lack of pericyte-specific markers and the lack of good cell culture systems for propagating these cells. In recent years the identification of a panel of markers that could be used to identify pericytes in vivo (NG2, PDGFβ-receptor, and CD146) has greatly facilitated their analysis in intact tissues [11–14], while at the same time improved cell culture methods have been devised to permit in vitro studies [15, 16]. Recently we described a novel yet simple approach of isolating and culturing highly pure, high density cultures of mouse brain pericytes [17]. What is more, these pericyte cultures can be maintained for extended periods of time at high purity making this a powerful system for analyzing pericyte behavior. In a recent study, we used this culture system to demonstrate that TNF-α promotes pericyte proliferation and migration [18], consistent with recent data that TNF-α stimulates cerebral pericyte migration and matrix metalloproteinase (MMP)-9 production [19]. Taken together, these observations support a fundamental role for TNF-α in mediating vascular remodeling. In this chapter, we briefly describe this culture system, as well as methods for examining different aspects of pericyte behavior, including cell adhesion, cell migration, and cell proliferation. These assays can be used to examine the influence of TNF-α or any factor of choice on pericyte behavior. In addition the assay methods of adhesion, migration, and proliferation can be applied to almost any cell type.

2 Materials

2.1 Cell Culture

2.1.1 Dissection Tools

1. Large scissors, straight.
2. Small scissors, curved.
3. Curved forceps.
4. Plastic 100 mm petri dishes, sterile, bacteriological grade (to contain the isolated brains).
5. 15 mL Polypropylene centrifuge tubes, sterile.
6. 70 % Ethanol in a spray bottle.

2.1.2 General Equipment

1. Biological safety cabinet.
2. Inverted microscope with ×10 and ×20 objectives.

3. A 37 °C incubator with humidity and gas control to maintain >95 % humidity and an atmosphere of 5 % CO_2 in air.
4. Pipetman.
5. Low-speed centrifuge.

2.1.3 Tissue Culture Tools

1. Tissue culture 6-well plates.
2. 15 and 50 mL polypropylene centrifuge tubes.
3. 10 mL plastic serological pipettes.
4. Plastic pipette tips: 1 mL and 200 μL.
5. 10 mL Syringes.
6. 19 Gauge needles.
7. 21 Gauge needles.

2.1.4 Media and Reagents

1. MEM-HEPES with penicillin/streptomycin.
2. Papain Cell Dissociation solution (Worthington Biochemical Corp., NJ). Add 5 mL of MEM-HEPES to the vial of papain and 0.5 mL of MEM-HEPES to the DNase tube.
3. 30 % Albumin solution from bovine serum.
4. Nutrient Mixture F12 Ham (Sigma-Aldrich), with sodium bicarbonate.
5. 100× penicillin/streptomycin solution. Aliquot into 5 mL lots and store at −20 °C. One aliquot of penicillin/streptomycin is added to 500 mL bottles of MEM-HEPES and Nutrient Mixture F12 Ham.
6. Endothelial cell growth supplement (Millipore 02-102): rehydrate 15 mg vial with 1 mL of Ham's F12 medium = 15 mg/mL stock. Aliquot into 100 μL aliquots and freeze at −20 °C. Add one aliquot (100 μL) to 50 mL base medium (Ham's F12 containing 10 % FBS and penicillin/streptomycin) to give working conc. = 30 μg/mL.
7. 400 mM L-glutamine solution: add 5 mL of F12 medium into the L-glutamine bottle, mix well (keep shaking the bottle during the aliquoting so as to keep the glutamine evenly distributed) and separate into 0.5 mL aliquots. Freeze at −20 °C (*see* **Note 1**). Add one aliquot to 50 mL base medium (1:100 dilution).
8. Ascorbate: add 10 mL of water to 5 mg ascorbate = 0.5 mg/mL stock, 0.25 mL aliquots, freeze at −20 °C, use at 1:200, add one aliquot to 50 mL base medium to give working conc. = 2.5 μg/mL.
9. Heparin: add 9 mL of F12 medium into a bottle containing 10,000 U (approx 72 mg) to give approx 0.8 % stock (8 mg/mL), separate into 0.25 mL aliquots, freeze at −20 °C. Add one 0.25 mL of aliquot to 50 mL base medium (1:200 dilution) to give working conc. = 40 μg/mL.

10. Stock F12 medium with penicillin/streptomycin: add 5 mL of aliquot of 10× penicillin/streptomycin to 500 mL Nutrient Mixture F12 Ham medium (with sodium bicarbonate).
11. Stock F12 medium with 10 % fetal bovine serum (FBS) and penicillin/streptomycin: add 50 mL of FBS aliquot and one 5 mL of 100× penicillin/streptomycin aliquot to 500 mL Nutrient Mixture F12 Ham medium (with sodium bicarbonate) to make stock F12 medium with FBS and antibiotics. FBS are stored in 50 mL aliquots at −20 °C (*see* **Note 2**).
12. Endothelial cell growth medium (ECGM): F12 medium with 10 % FBS and penicillin/streptomycin, endothelial cell growth supplement, ascorbate, L-glutamine, heparin. Aliquot 50 mL F12 stock medium with FBS and antibiotics into a 50 mL conical tube. Add 100 μL of endothelial cell growth supplement, 250 μL of ascorbate, 250 μL of heparin, 500 μL of L-glutamine (*see* **Note 3**).
13. Pericyte Growth Medium (PGM): Add 500 μL of Pericyte Growth Supplement (ScienCell, CA) into 50 mL of Pericyte Medium (ScienCell).
14. Collagen coating solution: collagen from calf skin Type I diluted 1:5 in sterile water.
15. 0.05 % Trypsin with EDTA.
16. Phosphate-buffered saline (PBS): 137 mM NaCl, 2.7 mM KCl, 8 mM Na_2HPO_4, 2 mM KH_2PO_4. Dilute 10× PBS to 1× PBS.

2.2 Cell Adhesion Assay

1. 24-Well tissue culture plates.
2. PBS.
3. Extracellular matrix proteins: collagen I, collagen IV, fibronectin, laminin, vitronectin (all from Sigma).
4. N1 serum-free medium: DMEM (high glucose + pyruvate), N1 supplement (Sigma), 4 mM L-glutamine, penicillin, streptomycin. To 50 mL of DMEM, add 0.5 mL of 100× N1 supplement, 0.5 mL of 100× penicillin and streptomycin, 0.5 mL of a 400 mM L-Glutamine solution (*see* **Note 2**).
5. Recombinant mouse TNF-α (R&D systems, MN).
6. 4 % Paraformaldehyde.

2.3 Cell Proliferation Assay

1. 24-Well tissue culture plates.
2. Microscope cover glass circles No. 1 to 0.13–0.17 mm thick; Size: 12 mm.
3. Extracellular matrix proteins: collagen I, collagen IV, fibronectin, laminin, vitronectin.
4. PBS.
5. N1 serum-free medium.

6. Recombinant mouse TNF-α.
7. Bromodeoxyuridine (BrdU) incorporation kit (Invitrogen Corp., CA).
8. Acid alcohol (5 % glacial acetic acid/95 % ethanol mixture) stored at −20 °C.
9. Platform for staining cover slips—we glue Eppendorf tube lids upside down on an inverted lid from a 100 mm Petri dish.
10. Curved forceps.
11. Wash tubes (50 mL) full of PBS.
12. Mouse anti-BrdU (from the BrdU kit).
13. Goat anti-mouse-FITC (from the BrdU kit).
14. Hoechst stain: Hoechst stain solution diluted 1:2 in PBS.
15. Aqua poly/mount (Polysciences).

2.4 Cell Migration Assay

1. 24-Well tissue culture plates.
2. Extracellular matrix proteins: collagen I, collagen IV, fibronectin, laminin, vitronectin.
3. PBS.
4. Pericyte Growth Medium.
5. N1 serum-free medium.
6. 1 mL Pipette tips.
7. Recombinant mouse TNF-α.

3 Methods

3.1 Isolation and Culture of Mouse Brain Pericytes

1. Make up fresh ECGM + FBS medium (*see* **Note 3**). Coat four wells of a 6-well plate with collagen type I diluted 1:5 in water (2 mL per well) for 2 h at 37 °C. Wash with PBS three times and keep the final wash in the well and transfer the plate to the incubator to ensure gas equilibration before adding the endothelial cells.
2. Prepare cultures from the brains of six mice (preferably 6–8 weeks old) euthanized by CO_2 inhalation (*see* **Notes 4** and **5**).
3. Flame all the tools and the razor blade to sterilize.
4. Ethanol spray the head and then use scissors to remove the skin on top of the skull. Next, use scissors to cut through the skull from the back of the head on both sides of the head towards the eyes. Carefully lift up the skull plate and remove the brain; place upside down in a 100 mm dish containing enough MEM-HEPES medium to totally cover the brain. Repeat this process for all six brains.

5. Transfer the brains to the lid of the 100 mm dish and then dice up the tissue using a flamed razor blade. Be thorough; chopping in both axes to ensure no tissue chunks remain.
6. Pour on some MEM-HEPES medium and aspirate the tissue using a 10 mL pipette into a 15 mL universal tube. Centrifuge this at 290 × *g* for 5 min.
7. At this stage, prepare the papain dissociation solution. Incubate these solutions in the 37 °C incubator for 10 min to ensure they have fully dissolved.
8. Aspirate the supernatant from the brain tissue and add 5 mL of papain solution and 250 μL of the DNase solution onto the brain tissue. Mix well and incubate at 37 °C for 1 h 10 min.
9. After 1 h and 10 min incubation, transfer the contents of the tube to a 50 mL tube to allow access of the syringe for trituration. Break up the tissue using a 10 mL syringe, first using a 19-gauge needle, ten times up and down, then with a 21-gauge needle. Be careful not to over triturate (*see* **Note 6**). The aim is to break up the microvessels from the brain tissue but to maintain the vascular tubes intact.
10. After trituration, transfer the brain homogenate (approximately 6–7 mL) to a new 15 mL tube and add approximately 7 mL of 22 % BSA to the tube and mix well. Centrifuge the 15 mL tube at 1,360 × *g* for 10 min. Following this spin, the myelin is retained at the top of the tube, the vascular tubes and liberated cells pellet at the bottom of the tube.
11. After the spin, tilt the tube at 45° to free the myelin from the edge of the tube, and pour the myelin plus BSA into a waster container. Remove the remainder of the myelin with a pipette, being careful not to allow the myelin to mix with the pellet. Suspend the blood vessel pellet with 1 mL of ECGM and transfer into a new 15 mL centrifuge tube. Add an additional 1 mL of ECGM medium to the suspension.
12. Wash the blood vessels by spinning down at 290 × *g* for 5 min.
13. Resuspend the cell pellet in 8 mL of ECGM and plate this onto four wells of a 6-well plate coated with type I collagen (*see* **step 1** in Subheading 3.1). Incubate in a 5 % CO_2, 37 °C, humidified incubator. Over the next few hours the vascular tubes attach and form colonies that continue to grow and expand with time.
14. Change all medium the next morning (day 1) replacing the ECGM with fresh ECGM and allow the cells to grow to confluence.
15. When the cultures are confluent, passage the cells. For the first passage, culture in ECGM, but from the second passage

onwards, switch to pericyte growth medium (PGM) instead of ECGM. Under these conditions, pericytes can be passaged multiple times to yield very pure pericyte cultures.

3.2 Cell Adhesion Assay

1. Prepare the ECM solutions to be coated onto plates. Collagen I is used at approximately 0.2 mg/mL, while all other ECM proteins are used at 10 μg/mL and diluted in PBS.
2. Coat the central areas of 24-well tissue culture plates with 25 μL drops of ECM solution (10 μg/mL) for 2 h at 37 °C (*see* **Notes** 7 and **8**).
3. Prepare the cells for the adhesion assay by harvesting and centrifuging the cells then resuspend the cells in a small volume of serum-free N1 medium (approximate volume = no. of drops × 25 μL).
4. Just before plating cells onto the ECM substrates, remove the ECM coating solutions, and wash the ECM-coated areas 2× with PBS or serum-free media (*see* **Note 9**).
5. Apply 25 μL of cell drops to the substrates, mixing the cell suspension well between the plating of each cell droplet.
6. Continue the adhesion assay for 15, 30, or 60 min to generate an adhesion time course.
7. After this time, stop the assay by adding 1 mL of PBS gently down the side of the wells, and gently swirl around the plate several times to remove unattached cells.
8. Remove the PBS and unattached cells and gently add 1 mL of 4 % paraformaldehyde at room temperature to fix the cells. After 20 min remove the paraformaldehyde and replace with 1 mL of PBS. Store the plates at 4 °C.
9. Quantify cell adhesion by counting the number of attached cells under light microscopy using a calibrated eyepiece graticule. For low density situations, count all cells in the droplet area. For high density situations, choose five random areas, by starting in the central zone, and then shifting to points North, East, South, and West (NESW) of this area.
10. Express the results as the number of cells attached (mean ± SD) for each condition. In the case of interventions such as effect of growth factors, or blocking antibodies or peptides, express the results as the percentage of cells attached under control conditions for each substrate.

3.3 Cell Proliferation Assay

1. Sterilize glass cover slips by dipping in 70 % ethanol and placing into wells of a 24-well plate. Allow to air dry in the tissue culture hood.
2. Prepare the ECM solutions to be coated onto the cover slips as described above (*see* Subheading 3.2).

3. Place 25 μL drops of ECM solution (10 μg/mL) into the center of the cover slips. Leave to coat for 2 h at 37 °C.
4. Prepare the cells by harvesting and centrifuging, then resuspend them in a small volume of media. Approximate volume = no. of drops × 25 μL.
5. Just before plating cells onto the ECM substrates, remove the ECM coating solutions, and wash the ECM-coated areas 1× with PBS.
6. Apply 25 μL of cell drops to the substrates, mixing the cell suspension well between the plating of each cell droplet.
7. Allow the cells to attach in the 37 °C incubator for 1 h. Under light microscopy, check that the majority of cells have attached and spread. If not, then extend this period for up to another hour.
8. Add 0.5 mL of medium to the well, and leave the cells to grow in the tissue culture incubator.
9. When the cells are approximately 50 % confluent (within the central region that they were plated in), add the BrdU labeling reagent to the tissue culture medium (1:500). The length of time of BrdU incubation depends on the basal rate of proliferation of the cell type under investigation (for primary brain pericytes incubate for 3 h) (*see* **Note 10**).
10. Remove the cover slips from the wells using the curved needle and curved forceps, and fix in acid/alcohol (95:5) for 20 min at −20 °C. This best achieved by placing the cover slips in a 24-well plate filled with acid/alcohol that is stored in the −20 °C freezer.
11. Wash the cover slips very well by dipping in PBS-filled 50 mL tubes. Take care to remove all acid/alcohol; any residual may destroy/denature the primary antibody and result in loss of signal (*see* **Note 11**).
12. Place the well-washed cover slips on the staining platform and add 50 μL drops of anti-BrdU antibody solution (anti-BrdU antibody diluted 1:50 in the specific incubation buffer from the kit). The anti-BrdU primary antibody vial is kept at −20 °C, and the specific incubation buffer in the fridge.
13. Incubate for 30 min at 37 °C.
14. Wash the cover slips well in PBS, and incubate with the anti-mouse-IgG-FITC solution (anti-mouse-FITC diluted 1:50 in PBS). Incubate for 30 min at 37 °C.
15. Wash the cover slips again, and incubate with Hoechst stain for 10 min at 37 °C.
16. Wash again and mount the cover slips with aquamount on a glass slide (maximum of five cover slips per slide). To remove

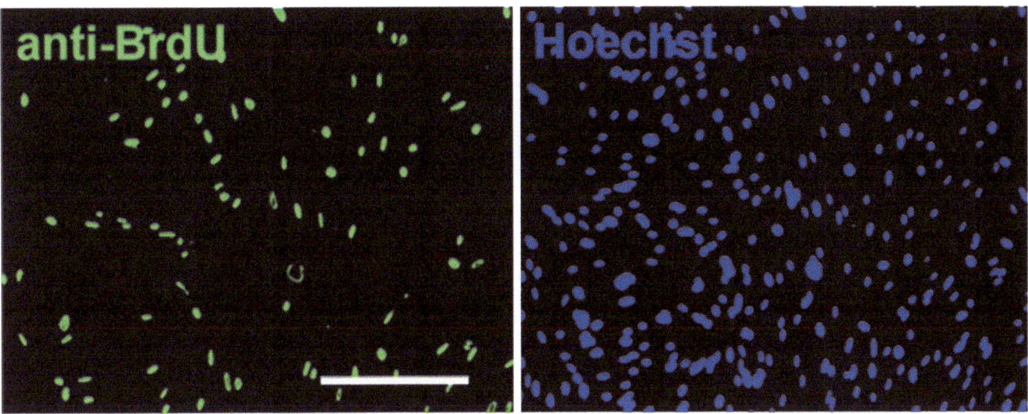

Fig. 1 The bromodeoxyuridine (BrdU) incorporation method of examining cell proliferation. Pericyte cultures were cultured on collagen I and stimulated with TNF-α for 12 h and cell proliferation was quantified using the BrdU incorporation method by immunofluorescent staining. Scale bar = 100 μm. The two panels show the same field of view stained with either an anti-BrdU antibody (to provide an index of the number of proliferating cells) or with Hoechst (to provide the total number of cells present)

excess aquamount, lay a piece of tissue paper on the top of the cover slips and apply gentle pressure to squeeze out the excess medium that is then absorbed by the tissue paper.

17. Store the cover slips in the dark until ready for analysis by fluorescent microscopy.
18. Examine under fluorescence. Quantify the total number of cells by Hoechst and the number of proliferating cells by BrdU-positive staining (Fig. 1). Randomly select several areas for quantification and calculate the means and standard deviations (SDs) for each condition.

3.4 Cell Migration Assay (Monolayer Scratch Assay)

1. Coat the wells of a 24-well tissue culture plate with 0.5 mL of ECM solution (10 μg/mL of each of the ECM proteins of interest) and incubate for 2 h in the 37 °C incubator.
2. Remove the ECM solution and wash the wells with 1 mL of PBS.
3. Prepare the pericytes by harvesting and centrifuging, then resuspend in PCM. Approximate volume = no. of wells × 0.5 mL.
4. Plate the pericytes onto the ECM-coated plates and culture until confluent.
5. After the cells have reached confluence (2–3 days after plating) it is time to make the scratch. Using a 1 mL pipette tip, scratch the monolayer by moving the tip across the monolayer from left to right to make a horizontal scratch. Then make the vertical scratch by rotating the plate through 90° and repeating this step. Check under the microscope that adequate scratches have been made.

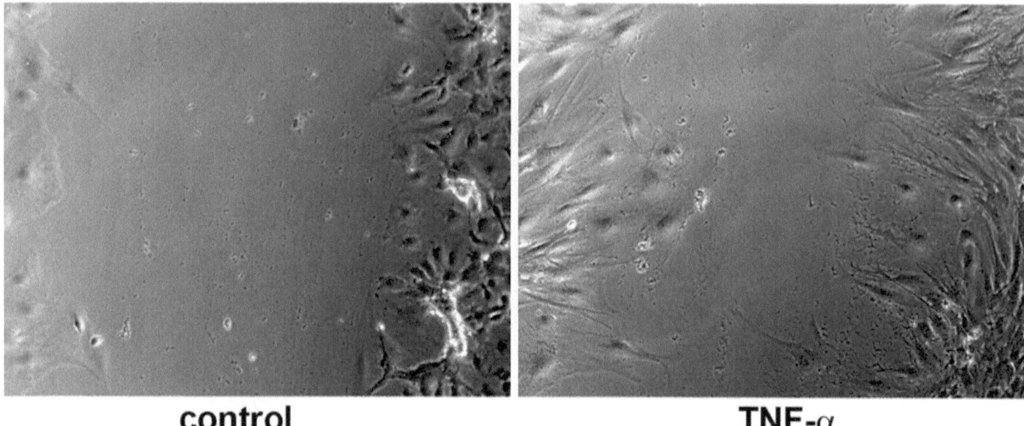

Fig. 2 TNF-α promotes pericyte migration in the migration scratch assay. Pericytes were cultured on collagen I. On reaching confluence, the pericyte monolayer was scratched to create bands that were devoid of cells. Over the next few hours cells migrate into the bare area and attempt to close the gap. 16 h after scratches were made, these phase pictures were taken. Scale bar = 100 μm. Note that TNF-α-enhanced pericyte migration, as illustrated by a more rapid closing of the scratch defect

6. Swirl the plate gently to move up the cellular debris and remove the medium and debris from the wells.
7. Replace with 1 mL of serum-free N1 medium per well.
8. Factors to be examined should be added at this time, e.g., cytokines, growth factors, or blocking antibodies or peptides.
9. Immediately after making the scratch, record the width of the scratch in all of the wells. Make 12 counts in each well. Start at the 12 o'clock position, measure nearest the cross-over point of the two scratches, then halfway along and near the end. Then repeat for the 3 o'clock, 6 o'clock, and 9 o'clock positions, to give a total of 12 counts for each well. Make these counts for all wells.
10. Return the plate to the incubator for 12–24 h. Deciding on when to stop the experiment depends on how fast the cells migrate (*see* **Note 12**) (Fig. 2).
11. At the end of the incubation period, make similar measurements (12 counts per well), and calculate the means and standard deviations of distance migrated (SDs).

4 Notes

1. When dissolving L-glutamine in small volumes of medium, it does not fully dissolve, but forms precipitates. Thus it is important to mix this concentrated solution very well and ensure all the L-glutamine is transferred to the 500 mL bottle, where it

fully dissolves in the larger volume. If dividing into 0.5 mL aliquots, shake the L-glutamine solution well in between each dispense to ensure equal distribution.

2. It is worth batch-testing several different sources of FBS. Most companies are happy to let you have free samples of their FBS to test. Once you have found the batch that supports the optimal growth of your cultures, stick with it and order several 500 mL bottles to store at −20 °C.
3. For every new isolation and culture, make up a fresh aliquot of 50 mL ECGM + FBS medium.
4. We find that it is important to dissect the brains from mice immediately after euthanization while mice are still warm—this gives the highest yield of cells.
5. We usually prepare brain vascular cultures with six brains going into four wells of a 6-well plate (1.5 brains per well). This protocol can be performed using fewer mice. However, it is critical that elements such as papain and DNase are downsized. For example, if using only one mouse brain (for instance when numbers of mice are limited such as with specific knockout strains), papain should be reduced to 833.3 μL and DNase to 41.7 μL and the cells should be plated into smaller sized wells, e.g., use wells of a 12-well plate.
6. When triturating the brain suspension, the aim is to break up the brain tissue with a slow methodical use of the syringe, but do not over triturate or move the tissue too violently as this will greatly reduce cell viability. BECs survive much better when they are contained within still intact vascular tubes; isolated BECs tend to die in the early stages of culture.
7. When moving the plates containing the small 25 μL drops around, take great care not to jolt or jar the plate, as this will move the droplet off the wetted area and will create a larger ECM-coated area, which will distort the size of the prepared area, and thus disrupt the experiment.
8. The use of plating 25 μL drops in the center of wells has several advantages over plating cells throughout the well. First, it uses fewer cells, which is an important factor when cell number is limited. Second, as the volume is much smaller (25 μL) than that needed to fill the entire well (0.5 mL), it is an effective way of conserving precious and expensive antibodies or peptides. Third, as all the cells are located within the central region of the cover slip and well, this optimizes visual acuity and makes it easy to observe changes in adhesion or morphology, as opposed to when cells are scattered throughout the well, many gather at the edge of the well, where visual optics are not optimal.
9. When washing the areas covered by the small 25 μL drops, add just 25 μL of wash volume, taking care not to move the droplet

off the wetted area, as this would create a larger area that the plated cells will occupy, thus distorting readouts of adhesion.

10. The length of time of BrdU incubation depends on the basal rate of proliferation of the cell type under investigation. For instance we find that pericytes have a relatively high basal rate of proliferation, and so 3 h BrdU incubation results in approximately 10 % of pericytes incorporating BrdU under non-stimulated conditions. TNF-α stimulation increases pericyte proliferation rates to 35 % [18]. In contrast to pericytes, primary brain endothelial cells (BEC) have a much lower basal proliferation rate [3], so require a longer BrdU incubation period to produce similar proliferation rates.
11. It is important to thoroughly remove all of the acid/alcohol fixative from the cover slip by washing several times in PBS-filled 50 mL tubes. If all the fixative is not removed, this will subsequently denature the added antibodies and prevent effective staining of cells. In our experience, this is the most common cause of failed immunofluorescence staining, particularly for the BrdU assay.
12. Deciding on when to stop the experiment depends on how fast the cells migrate. To show differences, it is ideal to wait until under those conditions showing most migration, the gap in the monolayer has almost been closed by the invading fronts of cells. The speed of migration depends on a number of factors including the cell type, cell density, age of cells, medium used, presence of cytokine or growth factor, or inhibitors.

Acknowledgements

This work was supported by the National Multiple Sclerosis Society; Harry Weaver Neuroscience Scholar Award (R.M.), and Postdoctoral Fellowship (J.V.W.-A.), by the American Heart Association (Western State Affiliate) Postdoctoral Fellowship (A.B.), and by the NIH grant RO1 NS060770.

References

1. Leibovich SJ, Polverini PJ, Shepard HM, Wiseman DM, Shively V, Nuseir N (1987) Macrophage-induced angiogenesis is mediated by tumour necrosis factor-α. Nature 329: 630–632
2. Baluk P, Yao LC, Feng J, Romano T, Jung SS, Schreiter JL et al (2009) TNF-α drives remodeling of blood vessels and lymphatics in sustained airway inflammation in mice. J Clin Invest 119:2954–2964
3. Welser J, Li L, Milner R (2010) Microglial activation state exerts a biphasic influence on brain endothelial cell proliferation by regulating the balance of TNF and TGF-β1. J Neuroinflammation 7:89
4. Hirschi KK, D'Amore PA (1996) Pericytes in the microvasculature. Cardiovasc Res 32:687–698
5. Bergers G, Song S (2005) The role of pericytes in blood-vessel formation and maintenance. Neuro Oncol 7:452–464

6. Dore-Duffy P, LaManna JC (2007) Physiologic angiodynamics in the brain. Antioxid Redox Signal 9:1363–1371
7. Hamilton NB, Attwell D, Hall CN (2010) Pericyte-mediated regulation of capillary diameter: a component of neurovascular coupling in health and disease. Front Neuroenergetics 2:5
8. Kutcher ME, Herman IM (2009) The pericyte: cellular regulator of microvascular blood flow. Microvasc Res 77:235–246
9. Herman IM, Jacobson S (1988) In situ analysis of microvascular pericytes in hypertensive rat brains. Tissue Cell 20:1–12
10. Hammes HP (2005) Pericytes and the pathogenesis of diabetic retinopathy. Horm Metab Res 37:39–43
11. Covas DT, Panepucci RA, Fontes AM, Silva WA Jr, Orellana MD, Freitas MC et al (2008) Multipotent mesenchymal stromal cells obtained from diverse human tissues share functional properties and gene-expression profile with CD146+ perivascular cells and fibroblasts. Exp Hematol 36:642–654
12. Li Q, Yu Y, Bischoff J, Mulliken JB, Olsen BR (2003) Differential expression of CD146 in tissues and endothelial cells derived from infantile haemangioma and normal human skin. J Pathol 201:296–302
13. Ozerdem U, Grako KA, Dahlin-Huppe K, Monosov E, Stallcup WB (2001) NG2 proteoglycan is expressed exclusively by mural cells during vascular morphogenesis. Dev Dyn 222:218–227
14. Sundberg C, Ljungstrom M, Lindmark G, Gerdin B, Rubin K (1993) Microvascular pericytes express platelet-derived growth factor-beta receptors in human healing wounds and colorectal adenocarcinoma. Am J Pathol 143: 1377–1388
15. Dore-Duffy P (2003) Isolation and characterization of cerebral microvascular pericytes. Methods Mol Med 89:375–382
16. Dore-Duffy P, Mehedi A, Wang X, Bradley M, Trotter R, Gow A (2011) Immortalized CNS pericytes are quiescent smooth muscle actin-negative and pluripotent. Microvasc Res 82: 18–27
17. Tigges U, Welser-Alves JV, Boroujerdi A, Milner R (2012) A novel and simple method for culturing pericytes from mouse brain. Microvasc Res 84:74–80
18. Tigges U, Boroujerdi A, Welser-Alves J, Milner R (2013) TNF-alpha promotes cerebral pericyte remodeling in vitro, via a switch from alpha1 to alpha2 integrins. J Neuroinflammation 10:33
19. Takata F, Dohgu S, Matsumoto J, Takahashi H, Machida T, Wakigawa T et al (2011) Brain pericytes among cells constituting the blood-brain barrier are highly sensitive to tumor necrosis factor-a, releasing matrix metalloproteinase-9 and migrating in vitro. J Neuroinflammation 8:106

Chapter 9

CD137 in Chronic Graft-Versus-Host Disease

Juyang Kim, Hong R. Cho, and Byungsuk Kwon

Abstract

Chronic graft-versus-host disease (GVHD) occurs in recipients of allogeneic hematopoietic stem cell transplantation with a high frequency. Preclinical animal chronic GVHD models outlined in this chapter allow for the delineation of events that occur during chronic GVHD development. The DBA/2 → (C56BL/6 × DBA/2) F1 (BDF1) model is characterized by systemic lupus erythematosus (SLE)-like phenotype. The B10.D2 → Balb/c model presents many features of autoimmune scleroderma. The former model is useful in defining how alloreactive donor $CD4^+$ T cells break B-cell tolerance, whereas the latter model is suitable for dissecting the pathogenesis of organ fibrosis. Our laboratory has demonstrated that injection of a single dose of strong CD137 agonists can prevent or cure chronic GVHD in these two models. In general, these models are particularly suited to screening the immunomodulatory therapeutics.

Key words Chronic graft-versus-host disease, CD137, Systemic lupus erythematosus, Autoimmune scleroderma, BDF1

1 Introduction

Allogeneic hematopoietic stem cell transplantation (HSCT) is the treatment of choice for hematological malignancies. However, mature T cells present in the transplant induce a high frequency of potentially lethal graft-versus-host disease (GVHD). GVHD is classified into two forms: acute and chronic, based on clinical manifestations and timing of disease occurrence. The predominant use of mobilized peripheral stem cells and non-myeloablative conditioning regimens have led to a decreased incidence rate of acute GVHD, but chronic GVHD remains highly debilitating conditions with high mortality [1]. Clinical presentations of chronic GVHD are highly variable and resemble autoimmune diseases, fibrosis and chronic inflammation of target organs [2]. Risk factors for chronic GVHD seem to be related to dysregulation of donor hematopoietic stem cell-derived T lymphocytes, which might be caused by thymic damage due to conditioning and/or acute GVHD, autoantibody production by B cells, and biased Th2 immune responses [1].

Jagadeesh Bayry (ed.), *The TNF Superfamily: Methods and Protocols*, Methods in Molecular Biology, vol. 1155, DOI 10.1007/978-1-4939-0669-7_9, © Springer Science+Business Media New York 2014

One of the difficulties in defining the pathophysiology of chronic GVHD is that preclinical animal models are limited and none of the animal models reflect all the features of clinical chronic GVHD [3]. Despite these limitations, the DBA/2 → (C57BL/6 × DBA/2) F1 (BDF1) and B10.D2 → Balb/c models are valuable in that each presents a phenotype of systemic lupus erythematosus (SLE) and autoimmune scleroderma, respectively. In the former model of chronic GVHD, parental DBA/2 $CD4^+$ T cells break host B-cell tolerance in unirradiated recipient mice, resulting in the production of autoantibodies directed against DNA and chromatin, and immune-complex glomerulonephritis. In contrast, infusion of T cells of the other parent, C57BL/6, induces acute GVHD. In these parent → F1 GVHD models, the fate of donor $CD8^+$ T cells (activation versus anergy) determines which type GVHD occurs in unirradiated host mice: $CD8^+$ T cells of the C57BL/6 strain attack host organs, including lymphoid organs, as a result of allostimulation by recipient antigen-presenting cells (APCs), while those of the DBA/2 strain rapidly fall into anergy and fail to deplete host B cells that have the potential to produce autoantibodies [4].

The parent → unirradiated F1 GVHD model is unique among GVHD models in that the development of GVHD is initiated only by alloreactivity under no inflammatory environment [5]. This property is useful in dissecting various issues related to alloimmune responses mediated by $CD4^+$ T cells. In the DBA/2 → BDF1 chronic GVHD model, haplo-identical host APCs directly stimulate the T-cell receptor (TCR) of donor $CD4^+$ T cells and once activated, they stimulate host B cells to proliferate and develop into autoreactive B cells. This model is utilized to define the roles of molecules involved in cognate interactions between donor $CD4^+$ T cells and host APCs or B cells, using blockers, stimulators, or mice deficient for a specific cell surface receptor. Another novel aspect of the DBA/2 → BDF1 model is its value in clarifying how the fate of donor $CD8^+$ T cells (cytotoxic T lymphocytes) determines the divergence of GVHD into an acute or chronic form or in defining how GVHD and graft-versus-leukemia (GVL) effects are segregated. For example, strong simulation with agonist anti-GITR mAb converts chronic GVHD towards an acute form in the DBA/2 → BDF1 model [4]. In the same model, agonist anti-CD40 mAb augments GVL effects without inducing acute GVHD, suggesting that the size of donor $CD8^+$ effector T cell pool is a determining factor for the segregation of GVL effects from acute GVHD [6]. Lastly, the DBA/2-into-BDF1 model is suitable for identifying potential therapeutics to treat SLE-like GVHD.

The B10.D2 → Balb/c chronic GVHD model is different from the parent → F1 chronic GVHD model in which MHC-matched but minor histocompatibility antigen (miHA)-mismatched donor T cells and bone marrow cells are transferred into the host conditioned with total body irradiation. The hallmark of this model is autoimmune scleroderma in which fibrotic changes of the skin and other organs are prominent. Antigen presentation of miHA in the context of H-2^d MHC class II molecules is critical in the evolution of chronic GVHD in mice [7]. This model has merits, as it shares many phenotypic features with the sclerodermatous form of clinical chronic GVHD [3]. Even though cutaneous GVHD has been well defined in this model, there are also fibrotic changes in other organs, including the gastrointestinal tract, liver, lung, and salivary glands. Fibrosis is well known to be mediated by Th2 immunity and thereby Th2 cells are believed to play a pivotal role in chronic GVHD. This model is valuable to researchers who are interested in multiple organ sclerosis.

CD137 is a member of the tumor necrosis factor (TNF) receptor superfamily that functions mainly as a strong co-stimulatory molecule for T cells [8]. However, recent studies have demonstrated a broad range of CD137 functions in non-T cells [9–11] and the revelation of critical roles of CD137L reverse signalling in inflammation adds another complication to the CD137 biology [12]. Deficiency of CD137 signalling in either donor $CD4^+$ T cells or $CD8^+$ T cells results in significantly reduced acute GVHD [13]. In contrast, more severe chronic GVHD occurs as a result of blockade of the CD137 signalling pathway in the DBA/2 → BDF1 model [14], as inactivated donor $CD8^+$ T cells are unable to delete host B cells [4, 14, 15]. Our group has demonstrated that agonist anti-CD137 mAb is highly effective in deleting alloreactive $CD4^+$ T and B cells, resulting in reversing chronic GVHD [16]. In addition, successful tests were performed in regards with the therapeutic efficacy of agonist anti-CD137 mAb in the scleroderma-like chronic GVHD model [17].

2 Materials

2.1 The DBA2 → BDF1 Chronic GVHD Model

2.1.1 Production of Agonist Anti-CD137 mAb

1. Hybridoma cell: 3H3 (Dr. Robert S. Mittler, Emory University of Medicine, Atlanta, GA).
2. Female Balb/c-nu/nu mice, 8–10 weeks old and pathogen-free.
3. 1-mL syringes and 30-G needles (BD Biosciences, San Jose, CA).
4. Pristine (Sigma-Aldrich, St. Louis, MO).
5. Trypan blue solution (Sigma-Aldrich).
6. Recombinant protein G agarose (Life Technologies, Grand Island, NY).
7. Poly-prep chromatography columns (Bio-Rad Laboratories, Hercules, CA).
8. Column stand.
9. 15- or 50-mL conical polypropylene centrifuge tubes (BD Biosciences).

10. Binding buffer: 0.53 g NaH_2PO_4, 0.87 g Na_2HPO_4, 0.877 g NaCl in 1 L of deionized water, pH 7.4.
11. Eluting buffer: 0.1 M glycine–HCl, pH 2.6.
12. Stripping buffer: 1 M acetic acid, pH 2.4.
13. Storage buffer: 0.05 % thimerosal in binding buffer.
14. Pierce™ BCA protein assay kit (Thermo Fisher Scientific, Inc., Rockford, IL).
15. Minisart syringe filter (0.2 μm pore size) (Sigma-Aldrich).
16. SnakeSkin® pleated dialysis tubing (3,500 MWCO) (Thermo Fisher Scientific, Inc.).
17. VIVASPIN 15R centrifugal concentrator (Sartorius AG, Goettingen, Germany).

2.1.2 Preparation of Donor Spleen and Lymph Node Cells and Induction of GVHD

1. Female DBA/2 (H-2^d) mice, 7–8 weeks old and pathogen-free.
2. 1-mL syringes and 30-G needles.
3. 15- or 50-mL conical polypropylene centrifuge tubes.
4. 60-mm-diameter tissue culture dishes (BD Biosciences).
5. Straight surgical scissors and forceps.
6. 40-μm-nylon mesh cell strainer.
7. Hemocytometer.
8. Heat lamp.
9. Gauze sponge.
10. Protective gloves.
11. Sterile PBS (phosphate-buffered saline, pH 7.4).
12. Erythrocyte hemolysis buffer: 144 mM NH_4Cl, 17 mM Tris–HCl, pH 7.2, in triple distilled water.

2.1.3 ELISA

1. 0.05 M carbonate/bicarbonate buffer: 1.59 g Na_2CO_3, 2.93 g $NaHCO_3$ in 1 L of distilled water, pH 9.6.
2. Coating buffer: 10 μg/mL of salmon sperm DNA (Sigma-Aldrich) in 0.05 M carbonate/bicarbonate buffer, pH 9.6.
3. Washing buffer: 0.1 % Tween 20 in PBS.
4. Blocking buffer: 2 % Bovine serum albumin (BSA) in PBS.
5. Horseradish peroxidase (HRP)-conjugated anti-mouse IgG1 (BD Biosciences).
6. Substrate: 1-Step™ Slow TMB-ELISA (Thermo Fisher Scientific, Inc.
7. Stop solution: 1 N HCl.
8. SpectraMax M2 microplate reader (Molecular Devices, Sunnyvale, CA).
9. Vacutainer blood collecting tube (BD Biosciences).

2.1.4 Histology and Immunohitochemistry

1. 10 % Formalin in PBS.
2. Hematoxylin and Eosin (H&E) solution.
3. Optimal cutting temperature (OCT) compound (Sakura Finetek, Tokyo, Japan).
4. Liquid nitrogen.
5. FITC-conjugated anti-mouse IgG (BD Biosciences).
6. Aluminum foil.
7. 15-mL conical tubes.
8. UltraCruz™ mounting medium (Santa Cruz Biotechnology, Inc., Dallas, TX).
9. Cryotome FSE cryostat microtome (Thermo Fisher Scientific).
10. Confocal microscope (Olympus, Tokyo, Japan).

2.1.5 Flow Cytometry

1. Staining buffer: PBS containing 0.2 % BSA and 0.1 % sodium azide.
2. Mouse Fc blocker: Purified rat anti-mouse CD16/CD32 mAb (2.4G2) (BD Biosciences).
3. PE-conjugated anti-mouse CD4 (GK1.5), PE-cy5-conjugated anti-mouse CD8a (53-6.7), FITC-conjugated anti-mouse CD62L (MEL-14), FITC-conjugated anti-mouse CD45R/B220 (RA3-6B2), PE-conjugated anti-mouse H-2K^b (AF6-88.5), and FITC-conjugated Annexin V (BD Biosciences).
4. FITC-conjugated rat IgG2a κ isotype control antibody (BD Biosciences).
5. 5-mL polystyrene round-bottom tube (BD Biosciences).
6. Vortex.
7. FACS Canto II (BD Bioscicnccs).

2.2 The B10.D2→Balb/c Chronic GVHD Model

2.2.1 Preparation of Donor Spleen and Lymph Node Cells and Induction GVHD

1. All materials described in Subheading 2.1.2.
2. B10.D2 (H-2^d) instead of C57BL/6 mice (*see* **Note 1**).
3. γ-Irradiator.
4. Antibiotics: Cotrim-ratiopharm (400 mg sulfamethoxazole, 80 mg trimethoprim in 5 mL ampullen).

2.2.2 Depletion of T Cells from Bone Marrow Cells and Purification of Donor CD4$^+$ T Cells

1. MACS® buffer: PBS containing 5 mM EDTA, 0.5 % BSA.
2. MACS® LS separation columns (Miltenyi Biotec, San Diego, CA).
3. Biotinylated anti-CD3 mAb (BD Biosciences).
4. Streptavidin-conjugated microbeads (Miltenyi Biotec).

5. Anti-CD4-conjugated magnetic beads (Miltenyi Biotec).
6. MACS® magnetic stand and magnets (Miltenyi Biotec).

2.2.3 Histology

1. All materials described in Subheading 2.1.4.
2. Masson's trichrome kit (Sigma-Aldrich).

3 Methods

3.1 The DBA2→ BDF1 Chronic GVHD Model (See Note 1)

3.1.1 Production of Ascites

1. Inject intraperitoneally 0.5 mL of pristine per mouse.
2. Inject intraperitoneally 5×10^6 of hybridoma cells 7–10 days after pristine priming.
3. Monitor mice for abdominal distension. It takes 10–14 days for mice to gain 25 % increase of body weight over the baseline established at inoculation. Harvest ascites by inserting a 18-G needle in the lower quadrant of the abdomen.
4. Clear ascites by centrifuging at $334 \times g$ for 10 min.
5. Store ascites fluid at −20 °C until the purification.

3.1.2 Purification of mAb

1. Equilibrate ascites with binding buffer until pH lies between 6.0 and 7.0.
2. Transfer 1 mL of recombinant protein G agarose slurry onto an empty column.
3. Wash the column with 10 mL of binding buffer.
4. Load 8–10 mL of ascites onto the column.
5. Wash the column with 8–10 mL of binding buffer to remove unbound protein.
6. Elute bound IgG with 6–10 mL of elution buffer. Immediately adjust pH of eluted IgG fractions to 7.0 with 1 M Tris base.
7. Reequilibrate the column with 10 mL of binding buffer. The column may be reused, starting from **step 4** or may be prepared for storage.
8. The column should be cleaned by washing with 5 mL of stripping buffer.
9. Equilibrate the column with 5 mL of storage buffer and store at 4 °C.
10. Dialyze the purified mAb against 4 L of PBS overnight at 4 °C.
11. Collect buffer-changed mAb.
12. Concentrate the antibody with VIVASPIN 15R concentrator. Centrifuge at $3,000 \times g$ at 4 °C.
13. Measure the antibody using the Pierce™ BCA protein assay kit.

3.1.3 Preparation of Lymphocytes from Spleen and Lymph Nodes

1. Euthanize DBA/2 mice by cervical dislocation.
2. Wash the mice with 70 % ethanol and place them in a laminar flow hood.
3. Open the abdomen and harvest the lymph nodes (cervical, axillary, brachial, inguinal, and mesenteric) and spleen into a petri dish that contain PBS. Most laboratories use only splenocytes as the source for mature donor T cells. However, our laboratory prefers both splenocytes and lymph node cells, because lymph nodes are enriched with mature T cells.
4. Homogenize lymph nodes and spleen by gently pressing them with the rubber tip of a sterile 1-mL syringe plunger.
5. Filter cells through cell strainer into a 50-mL polypropylene conical tube.
6. Centrifuge at 334 × *g* for 5 min.

3.1.4 Removal of Red Blood Cells from Lymph Node and Spleen Cell Suspensions

1. Resuspend harvested cells with 50 mL of erythrocyte hemolysis buffer to lyse erythrocytes.
2. Filter the remaining cells through a cell strainer.
3. Centrifuge at 334 × *g* at 4 °C for 5 min.
4. Resuspend cells in PBS and count viable cells using Trypan blue exclusion.

3.1.5 Induction of GVHD

1. Adjust the volume of the PBS in the tube to contain cells at the concentration of 4×10^8 cells/mL.
2. Warm mice with a heat lamp to dilate blood vessels.
3. Swab the tail with 70 % ethanol on a gauze sponge.
4. Inject 8×10^7 cells (200 μL) into BDF1 mice through the tail vein.

3.2 Assessment of SLE-Like GVHD

Chronic GVHD can be routinely assessed by measuring serum levels of IgG anti-DNA autoantibodies. IgG anti-DNA autoantibodies are detected in serum beginning from 2 weeks after disease induction. IgG_1 isotype is easily detected because of its high concentrations in serum. The severity of glomerulonephritis, a hallmark of SLE-like chronic GVHD, can be analyzed by indirectly measuring deposition of immune complexes in glomeruli using immunohistochemistry or by directly examining the degree of glomerular sclerosis from kidney sections.

Agonist anti-CD137 mAb is highly effective in preventing and reversing SLE-like GVHD. 200 μg of anti-CD137 mAb can be injected intraperitoneally immediately after disease induction to prevent the disease. The same protocol can be applied to treat advanced chronic GVHD. In this case, one-time injection between 30 and 100 days after disease induction is sufficient to reverse symptoms of the disease. Anti-CD137 mAb also has a therapeutic effect on advanced scleroderma-like chronic GVHD. In this model, however, it is important to note that anti-CD137 mAb shows a lethal toxicity when injected before 30 days after disease induction, as it exacerbates conditioning-induced systemic inflammation.

3.2.1 ELISA

1. Collect whole blood serum in a Vacutainer tube from the cut tail vein from 2 weeks after disease induction.
2. Allow the blood to clot by leaving it undisturbed at room temperature for 15–30 min.
3. Remove the clot by centrifuging at 1,000–2,000 × *g* at 4 °C for 10 min.
4. Transfer serum into a clean polypropylene tube and store it at −20 °C.
5. Coat a 96-well plate with 100 μL of carbonate buffer containing salmon sperm DNA at the concentration of 10 μg/mL overnight at 4 °C.
6. Wash the plate two times with washing buffer.
7. Add 100 μL of blocking buffer into each well and incubate at room temperature for 2 h.
8. Wash the plate two times with washing buffer.
9. Serially dilute serum samples in washing buffer.
10. Transfer 100 μL of the diluted samples into each well (in triplicate) and incubate at room temperature for 2 h.
11. Wash the plate two times with washing buffer.
12. Dilute detecting HRP-conjugated anti-mouse IgG1 at 1:4,000 in blocking buffer.
13. Add 100 μL of the diluted detecting antibody into each well.
14. Incubate at room temperature for 2 h.
15. Wash the plate two times with washing buffer.
16. Prepare substrate solution according to the manufacturer's protocol and transfer 100 μL of substrate solution into each well.
17. Incubate the plate at room temperature for 15–20 min.
18. Prepare stop solution and add 100 μL of stop solution into each well.
19. Read the plate at 450 nm in a dedicated microtiter plate spectrophotometer.

3.2.2 Renal Histology

1. Place harvested kidneys into 50-mL polystyrene round-bottom tube containing PBS.
2. Cut kidneys into a size suitable for a paraffin block.
3. Fix kidneys in 10 % formalin.

4. Embed fixed kidneys in paraffin.
5. Cut blocks by 4-μm thickness.
6. Stain sections with hematoxylin and eosin solution.

3.2.3 Immunohistochemistry

1. Dissect harvested kidneys and cut them into a size suitable for a base mold.
2. Make base mold by wrapping aluminum foil around a 15-mL conical tube.
3. Embed tissues in base molds filled with OCT.
4. Use forceps to hold base mold edge and immerge it into the liquid nitrogen. Hold until the tissue solidifies (around 30 s).
5. Remove tissues from liquid nitrogen and place blocked tissue on dry ice or −20 °C freezer.
6. Store frozen tissue blocks in −80 °C deep freezer for long-term storage.
7. For section, attach the frozen tissue block on the cryostat chuck. Allow the tissue block to equilibrate to the cryostat temperature (−20 °C) before cutting the sections. Cut 5–10 μm thick sections. Pick up sections onto slides.
8. Dry at room temperature till the sections are firmly attached to the slide.
9. Store the slides in −80 °C deep freezer until use.
10. For staining, fix the slides in 4 % paraformaldehyde for 1 h.
11. Wash slides in cold buffer for 5–10 min. Repeat wash.
12. Apply one to two drops of blocking buffer to tissue sections and allow them to remain in place for 10–20 min.
13. Tap off the blocking buffer.
14. Cover each section with diluted (usually 4,000 times) FITC-conjugated anti-mouse IgG at room temperature for 1 h.
15. Apply mount onto slides and seal it with a coverslip.
16. Examine the sections using a confocal microscope.

3.3 Assessment of Conversion of Chronic-to-Acute GVHD by Flow Cytometry

Under conditions in which activation of donor $CD8^+$ T cells is strongly induced, chronic GVHD is converted to acute GVHD in the DBA/2 → BDF1 model (*see* **Notes 1** and **2**). In this situation, activated donor $CD8^+$ T cells deplete hematopoietic cells in lymphoid organs, providing space of donor cell engraftment. A severe wasting syndrome that is routinely observed in mice with acute GVHD is rare in these mice but they frequently experience a severe lympho-depletion in the spleen and lymph nodes. Therefore, the most reliable parameter indicating conversion of chronic-to-acute GVHD is a decrease in splenocyte numbers and an increase in donor engraftment. Donor cells can be identified by staining splenocytes with anti-H-2^b antibody. H-2^b-negative cells are of donor origin.

1. Euthanize GVHD BDF1 mice by cervical dislocation.
2. Open the abdomen and harvest the spleen into a petri dish that contains PBS.
3. Homogenize the spleen by gently pressing them with the rubber tip of a sterile 1-mL syringe plunger.
4. Filter the homogenized spleen through a cell strainer into the 50-mL polypropylene conical tube.
5. Centrifuge at 334 × *g* for 5 min.
6. Resuspend the harvested cells with 50 mL of erythrocyte lysis buffer.
7. Filter the remaining cells through a 40-μm-nylon mesh.
8. Centrifuge at 334 × *g* at 4 °C for 5 min.
9. Resuspend the cells in PBS and count viable cells using Trypan blue exclusion.
10. Aliquot 5×10^5 cells into each tube.
11. Add 2.4G2 antibody to each tube, votex, and incubate at 4 °C for 10 min.
12. Wash cells two times with staining buffer.
13. Resuspend the cells in 100 μL of staining buffer.
14. Add 1 μg of FITC or PE-conjugated anti-H-$2K^b$, anti-CD4, anti-CD8, or anti-B220 mAb to each tube and incubate at 4 °C for 20 min.
15. Wash cells two times with staining buffer.
16. Resuspend the cells in 100 μL of staining buffer.
17. Go for acquisition of cells by flow cytometry.

3.4 The B10.D2 → Balb/c Model

In this model, transfer of total spleen and lymph node cells or purified donor $CD4^+$ T cells can result in chronic GVHD. As in the BDA/2 → BDF1 model, donor $CD4^+$ T cells alone are sufficient to induce this disease. To investigate the specific roles of donor $CD4^+$ T cells in chronic GVHD, GVHD can be induced by transfer of only purified donor $CD4^+$ T cells (*see* **Note 1**).

3.4.1 Preparation of Donor Spleen and Lymph Node Cells

Basically, we can adopt the same protocol for preparation of donor T cells as used in the DBA/2 → BDF1 model (*see* Subheading 3.1.3).

1. Prepare B10.D2 splenocytes and lymph node cells as described in Subheading 3.1.3.
2. Adjust the cell number to 1×10^7 in 90 μL of MACS® buffer and add 10 μL of anti-CD4-microbeads. After mixing well, incubate at 4 °C for 15–20 min.

3. Add 1–2 mL of MACS® buffer to the tube and centrifuge at 334×*g* for 5 min.
4. Remove the supernatant completely and resuspend the cell pellet in 500 μL of MACS® buffer.
5. Place the MACS® LS separation column in the magnetic filed.
6. Prepare the MACS® LS separation column by rising with 2 mL of MACS® buffer.
7. Apply cell suspension onto the column. Let unbound cells pass through. Rinse the column with 2 mL of MACS® buffer.
8. Remove the column from the magnetic field and place it onto a 15-mL tube. Add 2–3 mL of MACS® buffer to the column. Immediately flush out fraction that contains magnetically labelled cells.
9. Resuspend the cells in PBS and assess the viability of the cells using Trypan blue exclusion.

3.4.2 Depletion of T Cells from Bone Marrow Cells

1. Collect the bone marrow cells by flushing femurs and tibias from B10.D2 mice.
2. Centrifuge at 334×*g* for 5 min.
3. Resuspend the harvested cells with 50 mL of erythrocyte lysis buffer to lyse erythrocytes.
4. Filter the remaining cell through a 40-μm-nylon mesh cell strainer.
5. Centrifuge at 334×*g* at 4 °C for 5 min.
6. Resuspend cells in appropriate volume of PBS and count viable cells using Trypan blue exclusion.
7. Adjust the cell number to 1×10^7 in 90 μL of MACS® buffer and add 5 μL of biotinylated anti-CD3 mAb. After mixing well, incubate at 4 °C for 15–20 min.
8. Add 1–2 mL of MACS® buffer to the tube and centrifuge at 334×*g* for 5 min. Remove supernatant completely and resuspend cell pellet in 90 μL of MACS® buffer.
9. Add 5 μL of streptavidin-conjugated microbeads. After mixing well, incubate at 4 °C for 15–20 min.
10. Add 1–2 mL of MACS® buffer to the tube and centrifuge at 334×*g* for 5 min.
11. Remove the supernatant completely and resuspend the cell pellet in 500 μL of MACS® buffer.
12. Place the MACS® LS separation column in the magnetic filed.
13. Prepare the MACS® LS separation column by rising with 2 mL of MACS® buffer.
14. Apply cell suspension onto the column. Let unbound cells pass through.

15. Collect total effluent.
16. Wash the cells with PBS.

3.4.3 Induction of GVHD

1. Expose recipient Balb/c mice with 750 rad in a cesium irradiator.
2. Rest the irradiated mice for 3–4 h.
3. Inject 5×10^6 T-cell-depleted bone marrow cells and 5×10^7 spleen and lymph node cells in 200 μL of PBS, as described in Subheading 3.1.5. In case of purified donor $CD4^+$ T cells, inject 1×10^7 cells per mouse.
4. Add 5 mL of antibiotics (1 ample) to 500 mL of autoclaved drinking water and let mice drink ad libitum. Antibiotics are effective in preventing irradiated mice from dying due to infection.

3.4.4 Assessment of Scleroderma-Like GVHD

Skin lesions begin to appear 2 weeks after disease induction. This lesion includes alopecia and scaling on the ears, tail, and paws. Histological examination can further provide pathological features of skin lesions such as loss of dermal fat, hair follicle destruction, mononuclear cell infiltration, and increased collagen deposition.

1. Clinical scoring systems: The following scoring system is used—healthy appearance equals 0; skin lesions with alopecia less than 1 cm^2 in area, 1; skin lesions with alopecia 1–2 cm^2 in area, 2; skin lesions with alopecia more than 2 cm^2 in area, 3; additionally, animals are assigned 0.3 point each for skin disease (lesions or scaling) on the ear, tail, and paws (Minimum score 0 and maximum 3.9). Incidence rate can be calculated by counting mice that have score of 0.6 or higher.
2. Histology: Histological analysis can be performed as described in Subheadings 3.2.2 and 3.2.3. To clearly see collagen deposition, Masson's trichrome staining is recommended.

4 Notes

1. The DBA/2 → BDF1 model is most suitable for defining how alloreactivity breaks B-cell tolerance. However, this model has limitations due to a paucity of gene-manipulated mice available for studies. C57BL/6 congenic mice carrying H2-Ab1^{bm12} allele (BM12; mutation in an MHC class II gene) can be used either as a recipient or a donor to induce SLE-like chronic GVHD. Since most knockout mice with a C57BL/6 genetic background are easily available, the BM12 → C57BL/6 or C57BL/6 → BM12 chronic GVHD model is recommended to study roles of specific gene in SLE-like GVHD development. Note that acute GVHD occurs in irradiated recipients in this

genetic combination. It is not known whether autoreactive B cells are generated from donor hematopoietic stem cells in irradiated recipients. This issue has not been solved in the scleroderma-like GVHD model, either. Recently, the DBA/2 → sublethally irradiated Balb/c model has been shown to have a chronic GVHD phenotype of both autoantibody production and fibrotic changes [18]. This model is interesting in that donor B cells produce autoantibodies with help from donor $CD4^+$ T cells in a thymus-independent manner. Impaired thymic negative selection of $CD4^+$ T cells is known to result in the development of chronic GVHD with an autoimmune disease phenotype, including scleromerma [19]. Therefore, it seems that residual host immune cells survived by reduced conditioning play a critical role in autoantibody production by mature donor B cells and subsequent scleroderma caused by deposition of autoantibodies, whereas thymic damage by intense conditioning results in the generation of autoreactive $CD4^+$ T cells derived from donor hematopoietic stem cells. Therefore, we would like to recommend that researchers should consider the strength of conditioning in a chosen chronic GVHD model to determine which pathological phenotype is adequate for their study aims (SLE versus scleroderma versus both).

2. Stimulation or blockade of co-stimulatory pathways is a reliable strategy to enhance or reduce alloimmunity [5]. Agonist antibodies against co-stimulatory receptors are generally used to augment alloreactive T-cell functions. In contrast, antagonist antibodies to co-stimulatory molecules, or antibodies to their respective ligands or a fusion protein consisting of the Fc of IgG1 and their extracellular domain effectively interfere with receptor–ligand binding, thereby blocking their co-stimulatory signalling transduction. The efficacy of co-stimulatory blockers in the inhibition of alloresponses is highest when they are injected prior to the priming phase of T cells (two to three injections beginning at 1 day before disease induction are sufficient to inhibit alloreactive T-cell activation). Even though the in vivo efficacy of co-stimulatory blockers should be tested by titrating the dose, approximately 200 μg per mouse is a working dose in most cases. A similar protocol is applicable to co-stimulatory stimulators.

References

1. Blazar BR, Murphy WJ, Abedi M (2012) Advances in graft-versus-host disease biology and therapy. Nat Rev Immunol 12:443–458
2. Wolf D, von Lilienfeld-Toa M, Wolf AM, Schleuning M, von Bergwelt-Baildon M, Held SA et al (2012) Novel treatment concepts for graft-versus-host disease. Blood 119:16–25
3. Chu Y-W, Gress RE (2008) Murine models of chronic graft-versus-host disease: insights and unresolved issues. Biol Blood Marrow Transplant 14:365–378
4. Kim J, Choi WS, Kang H, Kim H-J, Suh J-H, Sakaguchi S et al (2006) Conversion of alloantigen-specific $CD8^+$ T cell anergy to

CD8+ T cell priming through in vivo ligation of glucocorticoid-induced TNF receptor. J Immunol 176:5223–5231

5. Kwon B (2010) Intervention with costimulatory pathway as a therapeutic approach for graft-versus-host disease. Exp Mol Med 42:675–683
6. Kim J, Park K, Kim HJ, Kim J, Kim HA, Jung D et al (2008) Breaking of CD8+ T cell tolerance through in vivo ligation of CD40 results in inhibition of chronic graft-versus-host disease and complete donor cell engraftment. J Immunol 181:7380–7389
7. Kaplan DH, Anderson BE, McNiff JM, Jain D, Schlomchik MJ, Shlomchik WD (2004) Target antigens determine graft-versus-host disease phenotype. J Immunol 173:5467–5475
8. Croft M (2009) The role of TNF superfamily members in T-cell function an diseases. Nat Rev Immunol 9:271–285
9. Lee SW, Park Y, Eun SY, Madireddi S, Cheroutre H, Croft M (2012) Cutting edge: 4-1BB controls regulatory activity in dendritic cells through promoting optimal expression of retinal dehydrogenase. J Immunol 189:2697–2701
10. Hsieh EH, Fernandez X, Wang J, Hamer M, Calvillo S, Croft M et al (2010) CD137 is required for M cell functional maturation but not lineage commitment. Am J Pathol 177: 666–676
11. Choi BK, Kim YH, Kwon PM, Lee SC, Kang SW, Kim MS et al (2009) 4-1BB functions as a survival factor in dendritic cells. J Immunol 182:4107–4115
12. Kim HJ, Lee JS, Kim JD, Cha HJ, Kim A, Lee SK et al (2012) Reverse signaling through the costimulatory ligand CD137L in epithelial cells is essential for natural killer cell-mediated acute tissue inflammation. Proc Natl Acad Sci U S A 109:E13–E22
13. Blazar BR, Kwon BS, Panoskaltsis-Mortari A, Kwak KB, Peschon JJ, Taylor PA (2001) Ligation of 4-1BB (CDw137) regulates graft-versus-host disease, graft-versus-leukemia, and graft rejection in allogeneic bone marrow transplant recipients. J Immunol 166:3174–3183
14. Nozawa K, Ohata J, Sakurai J, Hashimoto H, Miyajima H, Yagita H et al (2001) Preferential blockade of CD8(+) T cell responses by administration of anti-CD137 ligand monoclonal antibody results in differential effect on development of murine acute and chronic graft-versus-host diseases. J Immunol 167:4981–4986
15. Via CS (2010) Advances in lupus stemming from the parent-into-F1 model. Trends Immunol 31:236–245
16. Kim J, Choi WS, La S, Suh JH, Kim BS, Cho HR et al (2005) Stimulation with 4-1BB (CD137) inhibits chronic graft-versus-host disease by inducing activation-induced cell death of donor CD4 T cells. Blood 105:2206–2213
17. Kim J, Kim HJ, Park K, Kim J, Choi HJ, Yagita H et al (2007) Costimulatory molecule-targeted immunotherapy of cutaneous graft-versus-host disease. Blood 110:776–782
18. Zhang C, Todorov I, Zhang Z, Liu Y, Kandeel F, Forman S et al (2008) Donor CD4+ T and B cells in transplants induce chronic graft-versus-host disease with autoimmune manifestations. Blood 107:2993–3001
19. Sakoda Y, Hashimoto D, Asakura S, Takeuchi K, Harada M, Tanimoto M et al (2007) Donor-derived thymic-dependent T cells cause chronic graft-versus-host disease. Blood 1009: 1756–1764

Chapter 10

In Vitro Investigation of the Roles of the Proinflammatory Cytokines Tumor Necrosis Factor-α and Interleukin-1 in Murine Osteoclastogenesis

Joel Jules and Xu Feng

Abstract

Whereas the monocyte/macrophage-colony stimulating factor (M-CSF) and the receptor activator of NF-κB ligand (RANKL) are essential and sufficient for osteoclastogenesis, a number of other cytokines including two proinflammatory cytokines, tumor necrosis factor-α (TNF-α), and interleukin-1 (IL-1), can exert profound effects on the osteoclastogenic process. However, the precise mode of action of TNF-α and IL-1 in osteoclastogenesis remains controversial. While some groups demonstrated that these two cytokines can promote murine osteoclastogenesis in vitro in the presence of M-CSF only, we and others showed that TNF-α-/IL-1-mediated osteoclastogenesis requires permissive levels of RANKL. This chapter describes the method that we have used to investigate the effects of TNF-α and IL-1 on osteoclast formation in in vitro osteoclastogenesis assays using primary murine bone marrow macrophages (BMMs). Detailed experimental conditions are provided and critical points are discussed to help the reader use the method to independently evaluate the roles of TNF-α and IL-1 in osteoclastogenesis in vitro. Moreover, this method can be used to further elucidate the signaling mechanisms by which these two cytokines act in concert with RANKL or with each other to modulate osteoclastogenesis.

Key words Proinflammatory cytokine, TNF-α, IL-1, RANKL, RANK, Bone marrow macrophage, Osteoclasts, Osteoclastogenesis

Abbreviations

BMM	Bone marrow macrophage
FBS	Fetal bovine serum
IACUC	Institutional Animal Care and Use Committee
IL-1	Interleukin-1
LPS	Lipopolysaccharide
M-CSF	Macrophage/monocyte-colony stimulating factor
NFATc1	Nuclear factor of activated T cells, cytoplasmic 1
OPG	Osteoprotegerin
OSHA	Occupational Safety and Health Administration
PBS	Phosphate-buffered saline

Jagadeesh Bayry (ed.), *The TNF Superfamily: Methods and Protocols*, Methods in Molecular Biology, vol. 1155, DOI 10.1007/978-1-4939-0669-7_10, © Springer Science+Business Media New York 2014

PI	Principal investigator
RANK	Receptor activator of nuclear factor-κB
RANKL	Receptor-activator of nuclear factor-κB ligand
TNF-α	Tumor necrosis factor-α
TRAF	TNF receptor-associated factor

1 Introduction

Osteoclasts, the bone-resorbing cells, are multinucleated giant cells that play an indispensable role in bone homeostasis and repair [1] and are also implicated in bone loss associated with various pathological conditions including postmenopausal osteoporosis, periodontal disease, rheumatoid arthritis, and cancer bone metastasis [2]. Osteoclasts are hematopoietic in origin and differentiate from mononuclear hematopoietic cells of the monocyte/macrophage lineage [1]. Within the bone marrow, hematopoietic cytokines such as stem cell factor, interleukin (IL)-3, and IL-6 stimulate hematopoietic stem cells to generate common myeloid progenitors [3, 4]. Granulocyte/macrophage-colony stimulating factor (GM-CSF) then promotes differentiation of common myeloid progenitors into granulocyte/macrophage progenitors, which are further stimulated by monocyte/macrophage-colony stimulating factor (M-CSF) to give rise to cells of the monocyte/macrophage lineage, namely bone marrow macrophages (BMMs). Thus, primary BMMs isolated from mouse long bones are widely used as osteoclast precursors to generate murine osteoclasts in cell culture in vitro.

Early in vitro osteoclastogenesis assays mainly involved two methods: (a) the whole bone marrow culture system in which all bone marrow cells were used to assess osteoclast formation [5] and (b) the co-culture system in which BMMs were co-cultured with osteoblasts/stromal cells and stimulated with factors such as 1α, 25-dihydroxyvitamin D3 and dexamethasone to promote osteoclast formation [6]. While these methods contributed considerably to a better understanding of osteoclast biology, the presence of other cell types including osteoblasts/stromal cells often complicated the interpretation of the data. The cloning of RANKL has significantly simplified and improved in vitro osteoclast assays, permitting the investigation of the direct effects of cytokines on osteoclast precursors and/or mature osteoclasts. Moreover, the discovery of RANKL has helped reveal that osteoblasts/stromal cells in the whole bone morrow system and co-culture system support osteoclast differentiation primarily by serving as a source of RANKL as well as M-CSF.

M-CSF functions primarily in maintaining the proliferation and survival of osteoclast precursors, and RANKL is principally responsible for inducing the differentiation of these osteoclast precursors to osteoclasts. RANKL, a member of the TNF superfamily [7, 8], is expressed primarily by marrow stromal cells, osteoblasts, and osteocytes. RANKL exerts its effects by activating its receptor RANK, a member of the TNF receptor superfamily. RANK activation triggers the recruitment of specific TNF receptor-associated factors (TRAFs) through its cytoplasmic domain leading to the activation of six major signaling pathways: NF-κB, Akt, p38, ERK, JNK, and the nuclear factor of activated T cells, cytoplasmic 1 (NFATc1), a master regulator of osteoclast differentiation [1, 9]. Collectively, these activated signaling pathways participate in osteoclast differentiation, function, and survival in part through regulating the expression of numerous genes including those encoding matrix metalloproteinase 9 (MMP9), tartrate-resistant acid phosphatase (TRAP), carbonic anhydrase II (Car2), and cathepsin K (Ctsk) during the osteoclastogenic process [10, 11]. Nonetheless, RANKL has a decoy receptor called osteoprotegerin (OPG), which is also a member of the TNF receptor superfamily [12]. OPG inhibits RANKL functions and hence osteoclastogenesis by competing with RANK for binding RANKL.

TNF-α, which is also another TNF-superfamily member, shares many functional similarities with IL-1, a member of the IL-1 cytokine family. TNF-α and IL-1 are two proinflammatory cytokines that are produced by many cell types including fibroblasts, endothelial cells, stromal cells, and macrophages [13]. TNF-α and IL-1 play crucial roles in immune and inflammatory responses [14]. Importantly, TNF-α and IL-1 are also strong modulators of osteoclast formation and are implicated in bone loss associated with many inflammatory conditions [2]. TNF-α receptors are members of the TNF receptor family [15], and IL-1 receptors belong to the IL-1 receptor family [16]. Notably, TNF-α and IL-1 can also utilize TRAFs to activate most of the known osteoclast signaling pathways (NF-κB, p38, Akt, ERK, and JNK) through their respective receptors [15, 16].

Prior to the discovery of RANKL, investigation of the role of TNF-α and IL-1 in osteoclastogenesis was carried out by using the whole bone marrow and co-culture systems [17]. However, these two osteoclast formation assay systems are not suitable for addressing the direct effects of TNF-α and IL-1 on osteoclast precursors or mature osteoclasts.

The discovery of RANKL prompted bone biologists to examine the direct effects of cytokines on osteoclast precursors using the simplified method in which BMMs are stimulated with M-CSF and RANKL. Several initial studies showed that TNF-α induces the differentiation of BMMs into osteoclasts in the absence of RANKL [18, 19]. However, subsequent studies demonstrated that the two pro-inflammatory cytokines cannot promote osteoclastogenesis in the absence of RANKL [20–23]. In particular, one group also demonstrated that osteoclastogenesis induced by TNF-α under this condition in the absence of RANKL could be inhibited by OPG, indicating a role for RANKL in this assay [20]. Consistent with this finding, both TNF-α and IL-1 can induce RANKL expression by stromal cells [24, 25]. Taken together, these findings suggest that contamination of TNF-α/IL-1 osteoclastogenesis assays by other cell types such as osteoblasts or stromal cells may complicate these assays, leading to conflicting outcomes. Therefore, careful consideration should be taken when investigating the roles of TNF and IL-1 in osteoclast formation. Interestingly, we and others have also found that while TNF-α and IL-1 cannot promote osteoclastogenesis in the absence of RANKL, they can do so in the presence of permissive levels of RANKL [11, 20, 26, 27]. Moreover, we have further demonstrated that while TNF-α and IL-1 are unable to activate the expression of NFATc1, the four osteoclast genes (MMP9, Ctsk, TRAP, and Car2) in BMMs, RANKL renders NFATc1 and the osteoclast genes responsive to the two proinflammatory cytokines [11, 27].

Herein, we describe the method that we used in our studies to investigate the roles and signaling mechanisms of TNF-α and IL-1 in murine osteoclastogenesis using primary BMMs [11, 27]. The method comprises four major steps: (1) acquisition of bone marrow cells, (2) isolation and expansion of BMMs, (3) differentiation of BMMs to osteoclasts, and (4) characterization of osteoclasts. Although these steps appear to be routine procedures for in vitro osteoclast generation, we provide detailed experimental conditions and highlight critical points, which need special attention to ensure success in using this method. Given the controversy regarding the roles of TNF-α and IL-1 in osteoclastogenesis in vitro, this method may be used to independently address this discrepancy. Furthermore, this method may help in the efforts to delineate the molecular mechanisms by which IL-1 and TNF-α collaborate with RANKL or with each other to promote osteoclastogenesis. Finally, this method may be further modified to investigate the role of TNF-α and IL-1 in human osteoclastogenesis in vitro.

2 Materials

2.1 Mice

Our lab utilizes C3H and/or C57BL6 mice purchased from Harlan Industries for the TNF-α-/IL-1-mediated osteoclastogenesis assays. However, data should be reproducible with other mouse strains. Prior to performing primary osteoclastogenesis assays, principal investigators (PIs) should get the approval of the Institutional Animal Care and Use Committee (IACUC) from their institutions for their research procedures. IACUC, among other requirements, usually entails that rodents are humanely sacrificed to avoid unnecessary pain. Furthermore, PIs should assure that lab researchers who will be performing this method are cleared by the Occupational Safety and Health Administration (OSHA) of their institutions (*see* **Note 1**).

2.2 Cell Culture Equipments

Osteoclastogenesis assays require the use of standard cell culture equipment including

1. Biological safety cabinets.
2. 7 % CO_2 cell incubators at 37 °C.
3. Hemocytometers.
4. Inverted microscopes.
5. Water baths.
6. Centrifuges.
7. Syringes and needles.
8. Centrifuge/microcentrifuge tubes.
9. Dissection boards.
10. Culture plates.
11. Surgical tools including standard scissors, forceps, and, although not required, scalpels.

2.3 Chemical Solutions and Reagents

1. Fetal bovine sera (FBS) (*see* **Notes 2** and **3**).
2. Phosphate-buffered saline, pH 7.2 (PBS).
3. Ficoll Density Gradient solutions.
4. Leukocyte Acid Phosphatase kit (Sigma).
5. Formaldehyde.
6. Triton X-100.
7. Alexa Fluor 488-phalloidin.
8. Hoecsht 33258.

2.4 Culture Medium

1. Osteoclast Medium: alpha-minimum essential medium (α-MEM) supplemented with 1.98 g (per 1,000 mL) of sodium bicarbonate, 10 mL (1 %) of L-glutamine, 10 mL (1 %) of penicillin–streptomycin solution, 100 mL (10 %) of heat-inactivated FBS.
2. Master-medium solution A: Osteoclast medium, 44 ng/mL M-CSF, 10 ng/mL RANKL (1/10 of the optimal RANKL dose), 5 ng/mL TNF-α or 5 ng/mL IL-1.
3. Master-medium solution B (Positive control solution): Osteoclast medium, 44 ng/mL M-CSF, 100 ng/mL RANKL.
4. Master-medium solution C (Negative control solution): Osteoclast medium, 44 ng/mL M-CSF, 5 ng/mL TNF-α or 5 ng/mL IL-1.

2.5 Cytokine Acquisition (See Note 4)

1. RANKL.
2. M-CSF.
3. TNF-α.
4. IL-1.

3 Methods

The method is based on two recent research articles by our lab on the role of RANKL in osteoclastogenesis induced by TNF-α or IL-1 [11, 27] (*see* **Note 5**).

3.1 Bone Marrow Acquisition

1. Dissection tools (scissors, forceps, and scalpels) must be washed vigorously and sterilized by soaking in 70 % of ethanol in a 50 mL tube for at least 1 h before usage. The lab bench area, which will be utilized to dissect bones must also be cleaned with 70 % of ethanol at priori.
2. Prepare an osteoclast medium. Add 600 mL of distilled water to a bottle powder of α-MEM, as indicated by the manufacturer. Supplement the medium with other ingredients and make up the volume to 1,000 mL with distilled water. Mix the medium well, filter through a vacuum filter, and then maintain sterile. Medium is stored at 4 °C for up to 2 weeks and pre-warmed to 37 °C before usage.
3. Sacrifice mice in a humanely fashion as recommended by IACUC. Researchers can process 5–10 mice in a single preparation if needed.
4. Carry sacrificed mice to a flame sterilized bench area in the lab. Then, pin mice firmly onto a dissection board through the inferior and superior limbs with the ventral side facing up.
5. Douse mice with 70 % of ethanol to reduce bacterial contamination. If processing more than two mice, we suggest that researchers process two mice at the time.
6. Slice through the abdominal skin with scissors or scalpels by cutting the skin away from each pinned leg as to reveal the femurs. Then, pull away with forceps to expose the femurs and tibiae. Avoid dissecting through the arteries and abdominal cavity to prevent contamination (*see* **Note 6**).
7. Use scalpels/scissors and forceps to dissect through the knee joint then through the joint to release the femurs and tibiae. Remove all tissues from the bones and retain the bones in a sterile 100-mm culture plate on ice while processing other mice.
8. After processing all mice, attached a gauge needle to a 10-mL syringe and then load the syringe with osteoclast medium by pulling slowly on the plunger to avoid air bubbles.

9. Remove the ends of the bones with scissors and then flush the bone marrow by inserting the needle gently into the shaft of the bones and release the media slowly. We typically use 2.5 mL of medium per bone or 10 mL of medium per mice (two tibia and two femurs).
10. Culture marrow cells flushed into 10 mL of medium per mice onto 100 mm treated cell culture plates overnight in the cell incubator. This is to allow bone marrow stromal cells and other cells to attach on the plate while most hematopoietic progenitors are non-adherent.
11. Dissection tools should be washed vigorously and sterilized in 70 % of ethanol and then stored.

3.2 Isolation and Expansion of BMMs

1. On the following day, collect the non-adherent cells in a 50 mL tube. It is important to avoid collecting attached cells to reduce the risk of contamination by stromal cells (*see* **Note 7**). We always collect only 8 mL of the 10-mL cell medium containing the floated cells.
2. Add 3 mL of Ficoll Density Gradient solution per 5 mL of medium slowly to the bottom of the tube to avoid mixture. Ficoll Gradient Density solution sinks to the bottom of the tube and media moves up. Then, centrifuge at 30 × *g* with no brake and acceleration for 20 min at 4 °C.
3. After centrifugation, cells of monocyte/macrophage lineage appear white at the interphase of medium and Ficoll solution. Aspirate the solution on the top of the interphase gently and then collect the monocyte/macrophage cells in a sterilize 50 mL tube. Typically, we always collect 80 % of cells at the interphase to avoid contamination by stromal cells.
4. Wash cells twice with 10 mL of PBS and centrifuge at 380 × *g* for 10 min at 4 °C.
5. Count cells using the hemocytometer after suspending into 1 10 mL of medium. Next, dilute cells to a density of 50,000 cells per ml of medium. Supplement cell medium with 220 ng/mL of M-CSF and then mix cells very well but gently by pipetting a few times (*see* **Notes 8** and **9**).
6. Seed 50,000 cells per well of 24-well culture plate and culture for 48 h to allow cells to reach 40 % confluence. It is possible that additional timing may be required to reach that confluence. Therefore, cells should be monitored daily to avoid over confluence. We found 40 % of cell confluence to be ideal for subsequent osteoclast differentiation (*see* **Note 10**).

3.3 Differentiation of BMMs to Osteoclasts by TNF-α or IL-1 with RANKL Medium

RANKL can promote TNF-α- and IL-1-mediated osteoclastogenesis through either continuous treatment with permissive RANKL dosage or transient treatment with optimal RANKL dosage (*see* **Notes 11** and **12**).

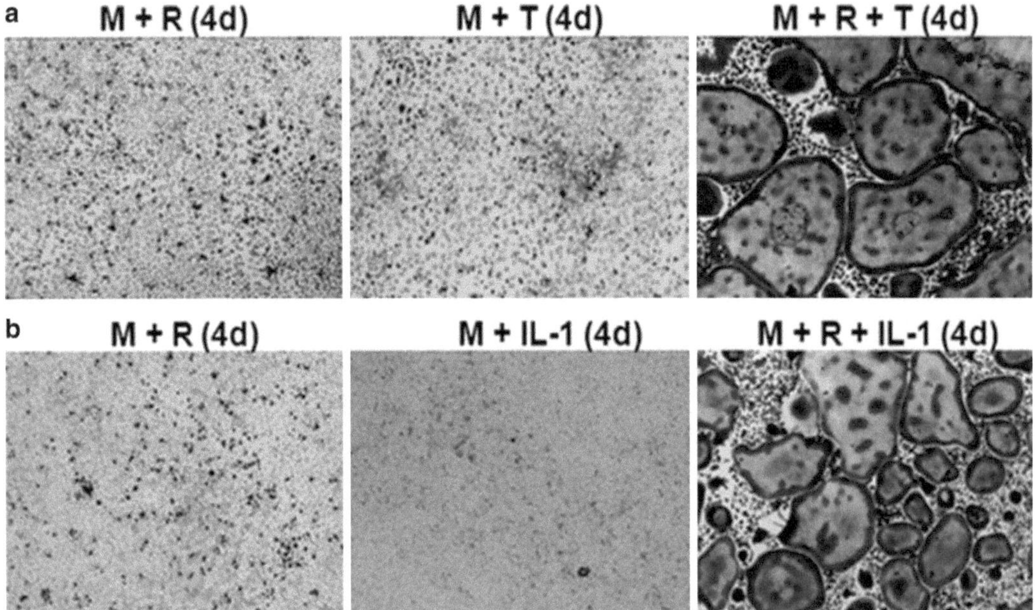

Fig. 1 Effect of permissive level of RANKL in TNF-α- and IL-1-mediated osteoclastogenesis. (**a** and **b**) 5×10^4 BMMs isolated from 4- to 6-week-old C3H mice were cultured on 24-well plates for 2 days (d) before treating with M-CSF (M, 44 ng/mL) and RANKL (R, 10 ng/mL) plus TNF-α (T, 5 ng/mL) (**a**) or IL-1 (5 ng/mL) (**b**) for 4 days. BMMs treated with 44 ng/mL M-CSF and 10 ng/mL RANKL, and 44 ng M-CSF plus 5 ng/mL TNF-α or 5 ng/mL IL-1 served as negative controls. Cells were then stained for TRAP activity on day 4. Results showed that TNF-α or IL-1 was unable to stimulate osteoclastogenesis with M-CSF only, but they could generate osteoclasts when attended by permissive RANKL dosage. Data indicate that TNF-α- and IL-1-mediated osteoclastogenesis require RANK-induced signaling

3.3.1 TNF-α- and IL-1-Mediated Osteoclastogenesis with Permissive RANKL

1. After the cells reach the correct confluence, prepare Master-medium solution A, Master-medium solution B (Positive control solution), and Master-medium solution C (Negative control solution) (Fig. 1) (*see* **Note 8**).
2. Mix the master solutions very well by pipetting a few times, each master solution will be added to three wells (1 mL/well) in 24-well culture plates, and then culture the cells for 4 days (*see* **Note 9**).
3. Medium supplemented with cytokines must be refreshed every 2 day depending on the duration of the experiment by gently aspirating the media and then replace it by fresh media.
4. It takes merely 4 days for BMMs to be differentiated to osteoclasts from TNF-α or IL-1 stimulation assisted by RANKL (Fig. 1). It may take longer dependently on the experimental conditions (*see* **Notes 13** and **14**).

3.3.2 TNF-α -and IL-1-Mediated Osteoclastogenesis with Optimal RANKL

1. After the cells reach the correct confluence, prepare Master-medium solution B (*see* **Note 10**). Mix the master solution very well, add 1 mL of media per well of 24-well culture plate, and then culture cells in the incubator for 18 h for the

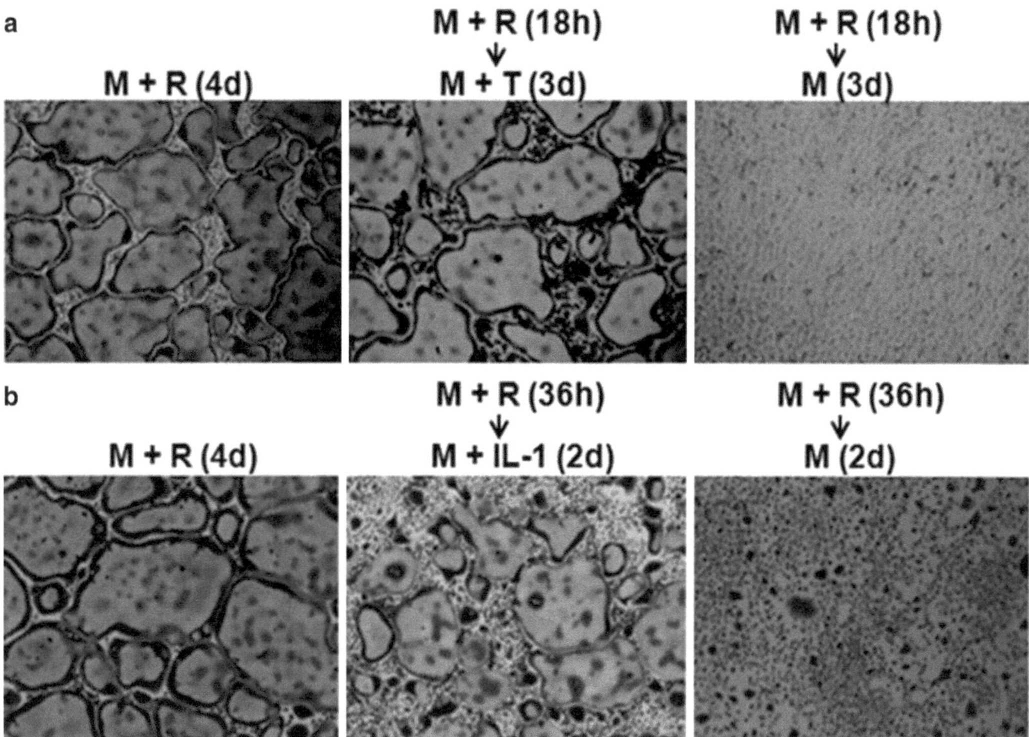

Fig. 2 Effect of RANKL priming in TNF-α- and IL-1-mediated osteoclastogenesis. (**a** and **b**) 5×10^4 BMMs isolated from 4- to 6-week-old C3H mice were cultured on 24-well plates for 2 days (d) before treating with M-CSF (M, 44 ng/mL) and RANKL (R, 100 ng/mL) for 18 or 36 h as indicated, washed with PBS, and then treated with 44 ng/mL M-CSF plus TNF-α (T, 5 ng/mL) (**a**) or IL-1 (5 ng/mL) (**b**) for 3 days or 2 days, respectively. BMMs treated with 44 ng/mL M-CSF and 100 ng/mL RANKL for 4 days served as positive controls, and BMMs primed with 44 ng/mL M-CSF and 100 ng/mL RANKL followed by treatment of M-CSF alone was used as negative controls. Cells were then stained for TRAP activity on day 4. Results showed that TNF-α or IL-1 could, similar to RANKL control, form osteoclasts in BMMs primed by RANKL. As negative control, BMMs primed by RANKL and treated with M-CSF alone generated no osteoclast. Data indicate that TNF-α- and IL-1-mediated osteoclastogenesis requires RANK-induced signaling

subsequent TNF-α stimulation or 36 h for the subsequent IL-1 stimulation (Fig. 2).

2. After the priming period, aspirate the medium gently then wash cells twice with PBS gently.
3. Prepare Master-medium solution C. Medium containing 44 ng/mL of M-CSF alone and Master-medium solution B would serve as negative and positive controls, respectively (*see* **Note 8**).
4. Mix the medium solutions very well, add 1 mL of media per 24-well plate, and culture cells for an additional 3 days for TNF-α stimulation or 2 days for IL-1 stimulation (*see* **Note 9**).
5. Medium supplemented with cytokines must be refreshed every 2 day dependent on the duration of experiment by gently aspirating the media and then replace it by fresh media.

6. It takes a total of 4 days for BMMs to differentiate to osteoclasts from TNF-α or IL-1 stimulation with the RANKL priming (Fig. 2). It may take longer depending on experimental conditions. But, it may not take longer than 6–8 days. The control assays should merely 4 days (*see* **Notes 13** and **14**).

3.4 Characterization of Primary Osteoclasts

At the end of the osteoclastogenesis assay, osteoclasts can be characterized by several techniques including TRAP staining, actin ring staining, and bone resorption.

3.4.1 TRAP Staining

1. Wash osteoclasts twice with PBS and then fix and stain them for TRAP activity with a Leukocyte Acid Phosphatase kit as indicated by the manufacturer.
2. Analyze and image stained osteoclasts with inverted microscope as shown in Figs. 1 and 2.
3. Osteoclasts can be quantified by counting multinucleated TRAP positive cells with three or more nuclei.

3.4.2 Actin Ring Staining

Osteoclasts can be generated on bone slices, which represent a physiological substratum for osteoclasts [11]. Subsequently, osteoclasts can be characterized for the formation of actin ring and multinucleation, two important features of mature osteoclasts.

1. Prepare BMMs as described under Subheadings 3.1 and 3.2 and then seed them on bone slices plated in 24-well plates. The preparation of bone slices is addressed in another chapter and will not be considered here.
2. Treat cells seeded on bone slices as discussed above under Subheading 3.3 with M-CSF plus TNF-α or IL-1 with 10 ng/mL of RANKL to generate osteoclasts with permissive RANKL levels for 4 days.
3. Alternatively, pretreat the cells with M-CSF and 100 ng/mL of RANKL, then wash them with PBS before culturing them with M-CSF plus TNF-α or IL-1 to form osteoclasts on the bone slices as discussed above.
4. At the end of the osteoclastogenesis assay, wash cells twice with PBS and then fix them with 3.7 % of formaldehyde solution in PBS for 10 min at room temperature.
5. After fixation, wash cells three times with PBS. Then, treat cells 0.1 % of Triton X-100 in PBS, a nonionic surfactant used to permeabilize the cell membrane, for 10 min to allow stain solution to get through the cells in the next step.
6. Stain cells with Alexa Fluor 488-phalloidin and Hoecsht 33258 for 15 min for actin ring and nucleic staining, respectively, as recommended by the manufacturer. Finally, analyze and image the bone slices using inverted confocal microscopy.

4 Notes

1. Biohazard Considerations. There are biohazard issues including injury, allergy, and zoonosis that are associated with working with mice in laboratory settings because researchers will be exposed to murine tissues and bone marrow, which may contain blood-borne pathogens. Therefore, researchers should be adequately trained in handling mice, isolating and processing murine tissues. Also, safety precautions should always and consistently be taken to minimize direct exposure and avoid splashing.
2. The choice of serum is an important factor and is available from many manufacturers. Serum supports cell growth by providing numerous essential macromolecules including attachment factors, nutrients, hormones, and growth factors to the cell culture medium. We find slight variations among different sources of serum and a number of lots of FBS are always tested carefully to ensure that they can support osteoclastogenesis in vitro prior to purchasing. It is thus very important for researchers to select specific sources of FBS that can work reproducibly well for their assays.
3. The osteoclast medium is supplemented with 10 % of heat-inactivated FBS. There are several unresolved controversies regarding the heat inactivation of FBS. FBS are usually inactivated to destroy labile components, such as complements, and bacterial contaminants, which can activate macrophages. There are reports suggesting that the levels of complement components found in FBS are either low or undetectable and also that bacterial contaminants are usually filtered from the serum and medium product, which would suggest that heat inactivation of FBS is unnecessary. There are further reports that extreme heat can affect macromolecules in FBS. Despite these reports, we still believe that heat inactivation of FBS is necessary for osteoclast differentiation. FBS stocks must be inactivated by heating at 56 °C for 30 min and then stored in aliquots (50–100 mL) at −20 °C.
4. As discussed above, the sources of recombinant RANKL, TNF-α, and IL-1 can have effects on this assay. They can be obtained from many manufacturers and maintained as recommended by the manufacturer. Among different potential factors that can cause this problem by the cytokines are potencies, kinetics, and endotoxin levels. Researchers are urged to carefully select their manufacturers because different commercial preparations of TNF-α, IL-1, and RANKL can vary significantly in dosing required for osteoclastogenesis assays. RANKL is reconstituted in 50–100 μg/mL with PBS in aliquots (25–50 μL) stored

at −80 °C as recommended by the manufacturer. A thawed RANKL aliquot can be stored at 4 °C if it is intended to be used up within several weeks. TNF-α and IL-1 are reconstituted in 10 μg/ml in aliquots of 20 μL and stored at −80 °C. A thawed aliquot can be kept at 4 °C if it is to be used up within several weeks.

5. We have only utilized this method in murine bone marrow cells to induce osteoclastogenesis and have not yet employed it in other TNF-α/IL-1 osteoclastogenesis assays by other species such as human cells. We therefore urge researchers to apply this method in other species.
6. Lipopolysaccharide (LPS) contamination can have drastic effects on osteoclastogenesis. We recently demonstrated that LPS could inhibit osteoclastogenesis in freshly isolated BMMs [28]. Our laboratory has always been very cautious about LPS contamination. We maintain special sets of glassware, graduated cylinders, flasks and bottles for osteoclast work. Moreover, we always pay special attention to endotoxin level when buying cytokines. We usually consider endotoxin level less than 0.01 EU per 1 μg of protein to be preferential.
7. Stromal-cell contamination is a serious problem that should be avoided. TNF-α and IL-1 can induce RANKL expression by stromal cells and osteoblasts which can affect osteoclastogenesis by these two cytokines. While the contribution of stromal cells in osteoclastogenesis induced by TNF-α/IL-1 with permissive RANKL levels may not be easily detected, the effect of stromal-cell contamination in TNF-α/IL-1 osteoclastogenesis by RANKL priming can be accessed because OPG administration can inhibit osteoclastogenesis resulted by RANKL induced by stromal-cell contamination but not osteoclastogenesis by TNF-α or IL-1. Our methodological approach was initially modified by culturing bone marrow cells overnight without M-CSF before isolating BMMs to minimize the risk of stromal-cell contamination. This approach is different from an earlier and still common approach where whole bone marrow cells are cultured for 3 days with M-CSF to expand bone marrow cells before collecting non-adherent cells [18]. The contamination of macrophages by stromal cells cannot be easily dismissed through this methodological approach as accessed by OPG-mediated inhibition of TNF-α-mediated osteoclastogenesis [20].
8. We prepared our mouse M-CSF from a M-CSF-producing cell line CMG14-12, which was generated and kindly, provided by Dr. Sunao Takeshita [29]. Our recombinant GST-RANKL was prepared as previously described [30]. In our osteoclastogenesis, we used 220 ng/mL of M-CSF to expand BMMs before osteoclast differentiation, and 10 ng/ml GST-RANKL plus

44 ng/mL M-CSF for TNF-α/IL-1-mediated osteoclastogenesis under the permissive condition, or 44 ng/ml M-CSF plus 100 ng/mL GST-RANKL to prime BMMs for subsequent osteoclastogenesis by TNF-α or IL-1. Nevertheless, osteoclastogenesis assays with highly purified and manufactured recombinant M-CSF and RANKL often require lower dosages. Medium containing 44 ng/mL of M-CSF alone can also be used as a negative control.

9. Macrophages can be easily activated, which can, in return, affect their abilities to differentiate to osteoclasts. It is important to handle these cells gently, especially when mixing and seeding them by avoid excessive pipetting.
10. Proper cell density is very important for osteoclast differentiation because a cell density, which is either too low or high, may delay osteoclastogenesis. We typically use 50,000 cells per well of 24-well plate that are first cultured for 48 h for cell expansion before submitting to treatments by osteoclastogenic factors for 4 days. We find that expanding BMMs to 40 % confluence before the osteoclast differentiation works ideally.
11. We find that the age of the mice can affect osteoclast formation by TNF-α and IL-1. The assay works best for young mice (4–6 weeks of age). Our data and others show that RANKL-mediated commitment of BMMs to the osteoclast lineage is a prerequisite for TNF-α- and IL-1-mediated osteoclastogenesis. Because RANKL levels in the bone marrow increase with age, we think that older mice may have BMMs that are already pre-committed by RANKL, which may thus affect the outcome of the assay.
12. Considering that RANK signaling is required for TNF-α- and IL-1-mediated osteoclastogenesis, the role of specific RANK signaling pathways in TNF-α-/IL-1-mediated osteoclastogenesis can be addressed by expressing specific RANK mutants in BMMs. Toward this end, bone marrow can be processed as discussed above but isolated BMMs can be seeded and cultured on 60 mm non-treated plates for 48 h for subsequent transfection. Thereafter, cells can be selected and then lifted with 2 mM EDTA in PBS for subsequent osteoclastogenesis as mentioned above.
13. In this method of TNF-α/IL-1-mediated osteoclastogenesis with RANKL, it takes 4 days to generate osteoclasts. Mononuclear TRAF positive cells appear on day 2 of the osteoclastogenic stimulation and then fuse to become multinucleated cells by day 4. Based on our experience, it is better to analyze and image osteoclasts on day 4 because their survival abilities tend to be a factor after 24 h of their formation. If unable to process immediately, cells can be fixed as recommended

by the manufacturer for subsequent TRAP staining (*see* Subheading 3.4.1) and stored at 4 °C for up to 1 week.

14. There are two types of IL-1: IL-1α and IL-1β [16, 31]. They share merely 26 % homology and bind to the same receptors (IL-1RI and IL-RII) but with different affinities. IL-1RI, the only receptor capable of mediating IL-1 action, binds IL-1α with greater affinity than it does IL-1β. IL-1RII, which functions mainly as a decoy receptor by antagonizing IL-1 action, binds stronger to IL-1β. Our assays use IL-1α since it is a stronger inducer of osteoclastogenesis than IL-1β.

Acknowledgments

This work is supported by National Institute of Arthritis and Musculoskeletal and Skin Diseases (NIAMS), National Institutes of Health grant AR47830 (to X. F.) NIAMS, National Institutes of Health Graduate Research Supplement to AR47830 (to J. J.), and by a *Within Our Reach* innovative basic research grant from the Research and Education Foundation of American College of Rheumatology (to X. F.).

References

1. Boyle WJ, Simonet WS, Lacey DL (2003) Osteoclast differentiation and activation. Nature 423:337–342
2. Feng X, McDonald JM (2011) Disorders of bone remodeling. Annu Rev Pathol 6:121–145
3. Kondo M, Wagers AJ, Manz MG, Prohaska SS, Scherer DC, Beilhack GF et al (2003) Biology of hematopoietic stem cells and progenitors: implications for clinical application. Annu Rev Immunol 21:759–806
4. Metcalf D (2008) Hematopoietic cytokines. Blood 111:485–491
5. van de Wijngaert FP, Tas MC, van der Meer JW, Burger EH (1987) Growth of osteoclast precursor-like cells from whole mouse bone marrow: inhibitory effect of CSF-1. Bone Miner 3:97–110
6. Udagawa N, Takahashi N, Akatsu T, Tanaka H, Tanaka H, Sasaki T, Nishihara T et al (1990) Origin of osteoclasts: mature monocytes and macrophages are capable of differentiating into osteoclasts under a suitable microenvironment prepared by bone marrow-derived stromal cells. Proc Natl Acad Sci U S A 87:7260–7264
7. Lacey DL, Timms E, Tan HL, Kelley MJ, Dunstan CR, Burgess T et al (1998) Osteoprotegerin ligand is a cytokine that regulates osteoclast differentiation and activation. Cell 93:165–176
8. Anderson DM, Maraskovsky E, Billingsley WL, Dougall WC, Tometsko ME, Roux ER et al (1997) A homologue of the TNF receptor and its ligand enhance T-cell growth and dendritic-cell function. Nature 390:175–179
9. Liu W, Xu D, Yang H, Xu H, Shi Z, Cao X et al (2004) Functional identification of three receptor activator of NF-kappa B cytoplasmic motifs mediating osteoclast differentiation and function. J Biol Chem 279:54759–54769
10. Rho J, Altmann CR, Socci ND, Merkov L, Kim N, So H et al (2002) Gene expression profiling of osteoclast differentiation by combined suppression subtractive hybridization (SSH) and cDNA microarray analysis. DNA Cell Biol 21:541–549
11. Jules J, Shi Z, Liu J, Xu D, Wang S, Feng X (2010) Receptor activator of NF-{kappa}B (RANK) cytoplasmic IVVY535-538 motif plays an essential role in tumor necrosis factor-{alpha} (TNF)-mediated osteoclastogenesis. J Biol Chem 285:37427–37435
12. Simonet WS, Lacey DL, Dunstan CR, Kelley M, Chang MS, Lüthy R et al (1997)

Osteoprotegerin: a novel secreted protein involved in the regulation of bone density. Cell 89:309–319

13. Vassalli P (1992) The pathophysiology of tumor necrosis factors. Annu Rev Immunol 10: 411–452
14. Kong YY, Feige U, Sarosi I, Bolon B, Tafuri A, Morony S et al (1999) Activated T cells regulate bone loss and joint destruction in adjuvant arthritis through osteoprotegerin ligand. Nature 402:304–309
15. Feng X (2005) Regulatory roles and molecular signaling of TNF family members in osteoclasts. Gene 350:1–13
16. Dinarello CA (2009) Immunological and inflammatory functions of the interleukin-1 family. Annu Rev Immunol 27:519–550
17. Takahashi N, Udagawa N, Kobayashi Y, Suda T (2007) Generation of osteoclasts in vitro, and assay of osteoclast activity. Methods Mol Med 135:285–301
18. Kobayashi K, Takahashi N, Jimi E, Udagawa N, Takami M, Kotake S et al (2000) Tumor necrosis factor alpha stimulates osteoclast differentiation by a mechanism independent of the ODF/RANKL-RANK interaction. J Exp Med 191:275–286
19. Azuma Y, Kaji K, Katogi R, Takeshita S, Kudo A (2000) Tumor necrosis factor-alpha induces differentiation of and bone resorption by osteoclasts. J Biol Chem 275:4858–4864
20. Lam J, Takeshita S, Barker JE, Kanagawa O, Ross FP, Teitelbaum SL (2000) TNF-alpha induces osteoclastogenesis by direct stimulation of macrophages exposed to permissive levels of RANK ligand. J Clin Invest 106:1481–1488
21. Li P, Schwarz EM, O'Keefe RJ, Ma L, Boyce BF, Xing L (2004) RANK signaling is not required for TNFalpha mediated increase in CD11(hi) osteoclast precursors but is essential for mature osteoclast formation in TNFalpha-mediated inflammatory arthritis. J Bone Miner Res 19:207–213
22. Ma T, Miyanishi K, Suen A, Epstein NJ, Tomita T, Smith RL et al (2004) Human interleukin-1-induced murine osteoclastogenesis is dependent on RANKL, but independent of TNF-alpha. Cytokine 26:138–144
23. Wei S, Kitaura H, Zhou P, Ross FP, Teitelbaum SL (2005) IL-1 mediates TNF-induced osteoclastogenesis. J Clin Invest 115:282–290
24. Quinn JM, Horwood NJ, Elliott J, Gillespie MT, Martin TJ (2000) Fibroblastic stromal cells express receptor activator of NF-kappa B ligand and support osteoclast differentiation. J Bone Miner Res 15:1459–1466
25. Kitaura H, Sands MS, Aya K, Zhou P, Hirayama T, Uthgenannt B et al (2004) Marrow stromal cells and osteoclast precursors differentially contribute to TNF-alpha-induced osteoclastogenesis in vivo. J Immunol 173:4838–4846
26. Li J, Sarosi I, Yan XQ, Morony S, Capparelli C, Tan HL et al (2000) RANK is the intrinsic hematopoietic cell surface receptor that controls osteoclastogenesis and regulation of bone mass and calcium metabolism. Proc Natl Acad Sci U S A 97:1566–1571
27. Jules J, Zhang P, Ashley JW, Wei S, Shi Z, Liu J et al (2012) Molecular basis of requirement of receptor activator of nuclear factor kappaB signaling for interleukin 1-mediated osteoclastogenesis. J Biol Chem 287:15728–15738
28. Liu J, Wang S, Zhang P, Said-Al-Naief N, Michalek SM, Feng X (2009) Molecular mechanism of the bifunctional role of lipopolysaccharide in osteoclastogenesis. J Biol Chem 284:12512–12523
29. Takeshita S, Kaji K, Kudo A (2000) Identification and characterization of the new osteoclast progenitor with macrophage phenotypes being able to differentiate into mature osteoclasts. J Bone Miner Res 15: 1477–1488
30. McHugh KP, Hodivala-Dilke K, Zheng MH, Namba N, Lam J, Novack D et al (2000) Mice lacking beta3 integrins are osteosclerotic because of dysfunctional osteoclasts. J Clin Invest 105:433–440
31. Dinarello CA (1984) Interleukin-1. Rev Infect Dis 6:51–95

Chapter 11

Evasion of TNF-α-Mediated Apoptosis by Hepatitis C Virus

Hangeun Kim and Ranjit Ray

Abstract

Hepatitis C virus (HCV) often causes chronic infection in humans, although the mechanisms for viral chronicity are not clearly understood. Tumor necrosis factor-α (TNF-α)-mediated apoptosis is a key element in a host organism's defense inhibiting viral spread and persistence. HCV has evolved mechanisms that antagonize host cell death signals so that virus propagation can continue unabated in infected cells. HCV core protein blocks TNF-α-mediated apoptosis signaling and inhibits caspase-8 activation by sustaining the expression of cellular FADD-like interleukin-1β-converting enzyme (FLICE)-like inhibitory protein (c-FLIP). HCV core protein also blocks TNF-induced proteolytic cleavage of the death substrate poly (SDP-ribose) polymerase from its native 116-kDa protein to the characteristic 85-kDa polypeptide. A decrease in endogenous c-FLIP by specific small-interfering RNA induces TNF-α-mediated apoptotic cell death and caspase-8 activation. However, HCV core neither affects the association between TNF receptor 1 (TNFR1) and TNFR1-associated death domain protein (TRADD) nor TRADD-Fas-associated death domain protein (FADD) and procaspase-8. Thus, HCV core protein appears to play a role in the inhibition of TNF-α-mediated cell death. This chapter describes methods to identify inhibitory mechanism of HCV for TNF-α-mediated apoptosis.

Key words Hepatitis C virus, Core protein, Tumor necrosis factor, Apoptosis, Caspases, Signaling mechanism, Human hepatoma cells

1 Introduction

Chronic HCV infection is a leading cause of progressive liver disease, including hepatic fibrosis, cirrhosis, and hepatocellular carcinoma. HCV efficiently escapes host immune responses and establishes chronic infection in >80 % infected humans. HCV infection induces interferon (IFN) but fails to clear virus. The protease inhibitors are used in association with peginterferon and ribavirin, which remain vital therapy components [1]. Protease inhibitor regimens substantially increased the sustained virological response (SVR) in naive patients and also in previously relapse patients. Because of the high risk of emerging resistance to protease inhibitors, prior null responders are not yet suitable candidates for triple therapy. The mechanisms responsible for HCV persistence are not well

Jagadeesh Bayry (ed.), *The TNF Superfamily: Methods and Protocols*, Methods in Molecular Biology, vol. 1155,
DOI 10.1007/978-1-4939-0669-7_11,

understood, although the interactions between HCV protein(s) and host cells appear to play an important role.

The pathology of HCV infection evolves around the development of a chronic, persistent infection with the potential for progression into hepatocellular carcinoma. Viral infection of host cells may induce an apoptotic response, and many viruses encode proteins which inhibit this mechanism. Apoptosis is the process of programmed cell death (PCD) that may occur in multicellular organisms, and it is a key element in the defense system of host organism against viral infections, viral spread, and viral persistence [2]. On the other hand, to circumvent host defense, viruses have evolved mechanisms that antagonize host cell death signals so that virus propagation can continue unabated in infected cells. Several mechanisms for the inhibition of apoptosis by viral proteins have been suggested. These include mimicking or inducing the Bcl-2 protein, blocking p53 function, inhibiting the activity of IL-1β-converting enzyme family members, or upregulation of insulin-like growth factor II [3–7].

HCV core protein [8] exhibits a gene regulatory role and the potential to suppress the onset of apoptotic cell death [9]. We previously investigated the signaling mechanism for inhibition of tumor necrosis factor-α (TNF-α)-mediated apoptosis after introduction of HCV core gene into human hepatoma cells. TNF activates both cell-survival and cell-death mechanisms simultaneously. Activation of NF-κB-dependent genes regulates the survival and proliferative effects of TNF, whereas activation of caspases regulates the apoptotic effects [10]. TNF superfamily members mediate both the intrinsic and extrinsic pathways of caspase activation. After binding of TNF to TNFR1, rapid recruitment of TRADD, RIP1, and TRAF2 occurs. Subsequently TNFR1, TRADD, and RIP1 become modified and dissociate from TNFR1. The liberated death domain (DD) of TRADD (and/or RIP1) binds to FADD, resulting in caspase-8/10 recruitment and apoptosis [11].

HCV core protein expression inhibited caspase-8 activation by sustaining the expression of cellular FADD-like interleukin-1β-converting enzyme (FLICE)-like inhibitory protein (c-FLIP) [12]. c-FLIP is a major resistance factor and critical anti-apoptotic regulator that inhibits TNF-α, Fas-L, and TNF-related apoptosis-inducing ligand (TRAIL)-induced apoptosis, as well as chemotherapy-triggered apoptosis in malignant cells [13]. Another group of investigators demonstrated that HCV NS5A protein interacts with TRAF2, which results in the inhibition of TNF-α-mediated NF-κB and JNK signaling activation [14, 15]. NS5B was also shown to interact with IKKα, which results in the inhibition of IKKα kinase activity. Thus, TNF-α-mediated NF-κB activation is blocked by HCV NS5B [16]. These data suggest that HCV inhibits TNF-α-mediated signaling pathway and may contribute to pathogenesis. The ability of HCV core protein to inhibit the TNF-α-mediated apoptotic signaling pathway may provide a selective advantage for virus replication, allowing for evasion of host antiviral defense mechanisms. Strategies to overcome c-FLIP-induced blockade of TNF-α-mediated apoptosis might be critical target for therapeutic intervention of chronic HCV infection.

2 Materials

1. Human recombinant TNF-α (Millipore; GF023).
2. Cell lines: Human breast adenocarcinoma cell line (MCF7), and liver hepatocellular carcinoma cell line (HepG2).
3. Dulbecco's modified Eagle medium.
4. Antibiotics: Puromycin and ampicillin.
5. cDNA clone Blue4/C5p-1 of strain HCV 1a containing the 5′ untranslated region, C, E1, E2, and a portion of the NS2 region (M. Houghton, Chiron Corp., Emeryville, USA).
6. *Eco*RI restriction enzyme (New England Biolabs, Inc., Ipswich, USA).
7. PCR reaction mix: 5 U Taq polymerase, 1× PCR buffer, 50 mM $MgCl_2$, 10 mM dNTP mixture, 10 μM each oligonucleotide primer (forward-5′-GTGCTTGCGAATTCCCCGGGA-3′, and reverse-5′-CGTGGAATTCGCACTTAGTAGG-3′, containing *Eco*RI restriction enzyme sites), and 1 μl cDNA clone in a total reaction volume of 50 μl.
8. pBabe/puro vector.
9. Ligation mix: 0.5 Weiss unit T4 DNA ligase, 1 μl 10× ligase buffer, 100 ng *Eco*RI-digested pBabe/puro vector, 300 ng *Eco*RI-digested core PCR product in a total reaction volume of 10 μl.
10. Calcium chloride competent *E. coli*.
11. SOC medium: 2 % bacto-tryptone, 0.5 % bacto yeast extract, 0.05 % NaCl, 20 mM glucose.
12. Lysogeny broth (LB) agar plates.
13. Lipofectamine reagent (Gibco BRL, Gaithersburg, USA).
14. HCV core sequence-specific primers: forward, 5′-GTGCTTG CGAATTCCCCGGGA-3′, reverse, 5′-CTTCCAGAATTCG GACGCCAT-3′.
15. Immunofluorescence microscopy.
16. Antibodies: Core-protein-specific antibody C7-50 (kindly provided by Dr. Jack R. Wands (Harvard Medical School, MA)); anti-FLAG antibody and anti-β-actin (Sigma-Aldrich, St. Louis, USA); anti-AU1 antibody (Abcam, Cambridge, USA); anti-caspase-3 monoclonal antibodies, anti-caspase-8

monoclonal antibodies, and anti-caspase-9 monoclonal antibodies (Cell signaling, Danvers, USA); anti-PARP monoclonal antibody (kindly provided by S. Chatterjee and N. Berger, Case Western Reserve University, Ohio); anti-mouse, anti-rabbit, and anti-goat immunoglobulin coupled with horse-radish peroxidase (Amersham, Pittsburgh, USA).

17. Cell Death Detection ELISAPLUS (Roche-applied-science, Indianapolis, USA) Caspase inhibitors: general caspase inhibitor Z-VAD-fmk (Bachem, Bubendorf, Sweden), caspase-8-specific inhibitor (Z-IETD-fmk, Calbiochem, San Diego, CA), caspase-9-specific inhibitor (Z-LEHD-fmk, Calbiochem), caspase-3-specific inhibitor (Z-DQMD-fmk, Calbiochem).
18. TNTG buffer: 30 mM Tris–HCl, pH 8.0, 150 mM NaCl, 1 % Triton X-100, 10 % glycerol.
19. Extraction buffer: 20 % SDS, 50 % dimethyl formamide, 2 % acetic acid, pH 4.7.
20. Multiscanner autoreader (Dynatech MR 5000, Dynatech Laboratories, Chantilly, VA).
21. 2× Laemmli reducing sample buffer: 126 mM Tris–HCl, pH 6.8, 20 % glycerol, 4 % SDS, 0.02 % Bromophenol blue.

3 Methods

3.1 Plasmid Construction and Generation of HCV Core Expressing Cell Lines

1. The cDNA clone Blue4/C5p-1 of strain HCV 1a containing the 5′ untranslated region, C, E1, E2, and a portion of the NS2 region was used in PCR to amplify the genomic region encoding the core protein (amino acids 1–191).
2. PCR reaction mix was used for a routine PCR amplification. Perform the amplification with an initial denaturation at 95 °C for 10 min, followed by 40 cycles at 95 °C for 30 s, 55 °C for 30 s, and 72 °C for 60 s, and finalized at 72 °C for 10 min.
3. After digestion using *Eco*RI restriction enzyme and gel purification, insert the amplified PCR product into the *Eco*RI site of the pBabe/puro vector under the control of murine leukemia virus long terminal repeat (MuLV) for expression of HCV core protein in cells.
4. Perform the ligation by incubating the ligation mixture at 16 °C for overnight.
5. Transform the ligation mixture to *E. coli* using calcium chloride transformation. Add ligation mixture to calcium chloride competent cells and incubated in ice for 30 min. Place the reaction tube at 42 °C water bath for 2 min and immediately transfer to ice for 5 min. After addition of 1 ml of SOC medium, incubate the transformed cells at 37 °C for 1 h, and plate

200 μL on ampicillin containing LB plate. Incubate the plates at 37 °C for overnight.

6. Select *E. coli* containing pBabe/puro-HCV 1a core plasmid construct and stock for further studies. Prepare pBabe/puro-HCV 1a core plasmid DNA after liquid culture of *E. coli* containing pBabe/puro-HCV 1a core for the next step.
7. Transfect subconfluent MCF7 cells (*see* **Note 1**) with pBabe/puro-HCV 1a core or control vector DNA using the Lipofectamine reagent.
8. Add 3 μg/mL of puromycin to cell culture medium to select and pool the puromycin-resistant stable transfectants.
9. Test the stable transfectants to confirm the integration of HCV core gene by nucleic acid analysis using routine PCR method with HCV core sequence-specific primers and protein expression using indirect immunofluorescence with HCV core-specific antibodies.

3.2 TNF-Mediated Apoptosis

Human cell lines expressing HCV core protein were used to study its potential to inhibit the apoptotic response. The apoptotic sensitivity of cells stably transfected with HCV core to TNF-α-mediated apoptosis was determined by Cell Death Detection ELISAPLUS kit that detects cytoplasmic histone-associated DNA fragments. Steps were performed according to the manufacturer's protocol.

1. Culture HepG2 cells (*see* **Note 1**) transfected with HCV core or empty vector control in 96-well plate (approximately 10^4 cells/well) for overnight, and then treat the cells with TNF-α (*see* **Note 2**) for 24–72 h and harvest.
2. Lyse the cells by incubation with 200 μL of lysis buffer (provided in supplier's kit) at room temperature for 30 min, followed by clarification at 200 × *g* for 10 min at room temperature.
3. Transfer 20 μL of clarified supernatant, containing DNA fragments onto the streptavidin-coated microplate.
4. Add 80 μL of Immunoreagent (anti-Histone biotin and anti-DNA POD) to each well, and incubate at room temperature for 2 h under gently shaking.
5. After removing the solution, rinse each well with 250 μl of incubation buffer (provided by the supplier in the kit).
6. Add 100 μL of ABTS (ready-to-use ELISA substrate for peroxidase driven indicator reactions) and incubate on a plate shaker at 20 × *g* until the color development is sufficient for a photometric analysis.
7. Detect the reaction in a spectrophotometry at 405 nm.
8. In some experiments, apply caspase inhibitors at the concentration of 20 μM to the medium 30 min before treatment.

3.3 Assessment of TNF-Cytotoxicity in MCF7 Cells

1. Expose approximately 5×10^4 MCF cells to 15 ng/mL of TNF-α (*see* **Note 3**) for 18 h in the absence of cycloheximide and incubate for an additional 48 h in TNF-α-free medium.
2. Determine the sensitivity of MCF7 cells to TNF-α in the absence of cycloheximide by the modified tetrazolium salt (MTT) assay.
   (a) Incubate the cells for indicated time frame in the presence or absence of different concentrations of TNF-α in a final volume of 0.2 mL for 72 h.
   (b) Determine cell viability by the addition of 0.02 mL of a 5 mg/mL solution of MTT.
   (c) After 2 h of incubation at 37 °C, add 0.1 mL of extraction buffer.
   (d) After incubation overnight at 37 °C, measure the optical density at 590 nm using a 96-well multiscanner autoreader with the extraction buffer serving as a blank.

3.4 In Vitro Co-immunoprecipitation Assay

1. To understand the potential role of core protein in disruption of the interaction between TNFR1-TRADD, transfect the HepG2 cells with a TNFR1 construct containing an N-terminal FLAG-tagged region (FLAG-TNFR1) and a C-terminal hemagglutinin (HA)-TRADD construct (TRADD-HA) using Lipofectamine.
2. Use empty-vector DNA in the place of core plasmid DNA as a control.
3. Also, transfect HepG2 cells with TRADD-HA and AU1-tagged FADD (FADD-AU1) in the presence or absence of a FLAG-tagged core construct (FLAG-core) to determine if core protein interrupts TRADD–FADD–procaspase-8 interaction.
4. After 24 h incubation, lyse the transfectants with TNTG buffer supplemented with a cocktail of protease inhibitors.
5. After sonication, precipitate the cell debris by centrifugation at a maximum speed for 30 min, and transfer clear cell lysates into fresh microtube.
6. Add anti-FLAG (for TNFR1–TRADD interaction) or anti-AU1 antibody (for TRADD–FADD–procaspase-8 interaction) and protein G-Sepharose agarose beads to the cell lysates and mix for 14 h.
7. Wash the beads four times with TNFTG buffer and subject to SDS-PAGE for Western blotting.

3.5 Detection of Caspase-3, Caspase-8, and Caspase-9 Activation

1. Treat hepG2 cells transfected with HCV core DNA plasmid or empty control vector with or without TNF-α.
2. Lyse the cells with 2× Laemmli reducing sample buffer. Run cell lysates on SDS-PAGE and transfer the resolved proteins onto nitrocellulose membrane.

3. Detect the cleavage of procaspase-3, procaspase-8, and procaspase-9 with specific monoclonal antibodies by Western blot analysis.
4. Use anti-β-actin monoclonal antibody for loading control.

3.6 Poly (ADP-Ribose) Polymerase (PARP) Cleavage Assay

1. Stimulate MCF7 cells stably transfected with or without HCV core protein with TNF-α.
2. Transfer similar amounts of cellular proteins onto nitrocellulose membrane after SDS-PAGE.
3. Use anti-PARP monoclonal antibody at a dilution of 1:500.
4. Use a mouse secondary antibody conjugated with horseradish peroxidase at a dilution of 1:10,000.
5. Visualize the peroxidase signal by chemiluminescence (ECL).
6. Estimate the PARP cleavage products from the migration of standard protein molecular weight markers.

4 Notes

1. MCF7 and HepG2 cell lines and their transfected derivatives were maintained in Dulbecco's modified Eagle medium containing 10 % fetal calf serum and a lower dose of the selection antibiotic (1 μg of puromycin/ml or 400 μg of G418/mL). Cells were incubated in a humidified incubator of 5 % CO_2 at 37 °C.
2. Human recombinant TNF-α is a potent lymphoid factor that exerts cytotoxic effects on a wide range of tumor cells and certain other target cells. The mature form of human TNF-α has 157 amino acid residues (17 kDa). Millipore GF023 is a recombinant polypeptide containing the mature form of human TNF-α with a polyhistidine tag at its amino terminal. The source of TNF-α is *E. coli* and lyophilized from 3 mM Tris, pH 8.0. The lyophilized powder is stable at room temperature and reconstituted with sterile distilled water to a concentration of 0.1–1 mg/mL. Reconstituted TNF-α should be stored in undiluted aliquots at −20 °C. TNF-α exerts its biological activity in the concentration range of 0.05–20 ng/mL.
3. The level of TNF-α in the chronic HCV patients was higher than healthy donors and was positively correlated with the severity of the disease. However, higher level of TNF-α failed to induce apoptosis or virus clearance in the therapeutic trials using interferon-α [17]. Our findings may explain how HCV inhibits TNF-α-mediated apoptosis and establishes persistent infection. Further studies may help to clarify the interference of HCV with apoptotic pathways and its role in the pathogenesis.

Acknowledgements

Our research was supported by grants R01CA85486, RO1DK 080812 and Internal Blue Ribbon Funds from Saint Louis University, Missouri.

References

1. Popescu C, Gliga S, Aramă V (2012) Trends in hepatitis C virus infection therapy: protease inhibitors a step forward in the era of direct acting antivirals. Rom J Intern Med 50:117–127
2. Shen Y, Shenk TE (1995) Viruses and apoptosis. Curr Opin Genet Dev 5:105–111
3. Subramanian T, Boyd JM, Chinnadurai G (1995) Functional substitution identifies a cell survival promoting domain common to adenovirus E1B 19kDa and Bcl-2 proteins. Oncogene 11:2403–2409
4. Wang XW, Gibson MK, Vermeulen W, Yeh H, Forrester K, Stürzbecher HW et al (1995) Abrogation of p53-induced apoptosis by the hepatitis B virus X gene. Cancer Res 55:6012–6016
5. Ueda K, Ganem D (1996) Apoptosis is induced by N-myc expression in hepatocytes, a frequent event in hepadnavirus oncogenesis, and is blocked by insulin-like growth factor II. J Virol 70:1375–1383
6. Vaux DL, Strasser A (1996) The molecular biology of apoptosis. Proc Natl Acad Sci U S A 93:2239–2244
7. White E (1996) Life, death, and the pursuit of apoptosis. Genes Dev 10:1–15
8. Lagging LM, Meyer K, Hoft D, Houghton M, Belshe RB, Ray R (1995) Immune responses to plasmid DNA encoding the hepatitis C virus core protein. J Virol 69:5859–5863
9. Ray RB, Meyer K, Ray R (1996) Suppression of apoptotic cell death by hepatitis C virus core protein. Virology 226:176–182
10. Rath PC, Aggarwal BB (1999) TNF-induced signaling in apoptosis. J Clin Immunol 19: 350–364
11. Micheau O, Tschopp J (2003) Induction of TNF receptor I-mediated apoptosis via two sequential signaling complexes. Cell 114: 181–190
12. Saito K, Meyer K, Warner R, Basu A, Ray RB, Ray R (2006) Hepatitis C virus core protein inhibits tumor necrosis factor alpha-mediated apoptosis by a protective effect involving cellular FLICE inhibitory protein. J Virol 80: 4372–4379
13. Safa AR, Pollok KE (2011) Targeting the anti-apoptotic protein c-FLIP for cancer therapy. Cancers (Basel) 3:1639–1671
14. Park KJ, Choi SH, Lee SY, Hwang SB, Lai MM (2002) Nonstructural 5A protein of hepatitis C virus modulates tumor necrosis factor alpha-stimulated nuclear factor kappa B activation. J Biol Chem 277: 13122–13128
15. Park KJ, Choi SH, Choi DH, Park JM, Yie SW, Lee SY et al (2003) Hepatitis C virus NS5A protein modulates c-Jun N-terminal kinase through interaction with tumor necrosis factor receptor-associated factor 2. J Biol Chem 278:30711–30718
16. Choi SH, Park KJ, Ahn BY, Jung G, Lai MM, Hwang SB (2006) Hepatitis C virus non-structural 5B protein regulates tumor necrosis factor alpha signaling through effects on cellular IkappaB kinase. Mol Cell Biol 26:3048–3059
17. Larrea E, Garcia N, Qian C, Civeira MP, Prieto J (1996) Tumor necrosis factor alpha gene expression and the response to interferon in chronic hepatitis C. Hepatology 23:210–217

Chapter 12

TNF-α Modulates TLR2-Dependent Responses During Mycobacterial Infection

Sahana Holla, Jamma Trinath, and Kithiganahalli Narayanaswamy Balaji

Abstract

Multifunctional roles of tumor necrosis factor-alpha (TNF-α) during the mycobacterial pathogenesis make it an important molecule to understand and to examine the course of infection. Identification and analysis of TNF-α response can largely contribute to determine the potential host mediators for therapeutic intervention against tuberculosis. The current chapter describes several methods to assess the ability of TNF-α signaling to modulate toll-like receptor (TLR)2 signaling, another key player in mycobacterial infection and its responses. Experiments involving neutralizing antibodies, antagonists, pharmacological inhibitors, and siRNA-mediated gene silencing are discussed in this chapter to establish the role of TNF-α signaling. The widely used protein and mRNA analysis readouts like enzyme-linked immunosorbent assay (ELISA), immunoblotting, fluorescence-activated cell sorting (FACS), and quantitative real-time RT-PCR are useful to estimate and confirm the mediators involved in TNF-α and TLR2 signaling.

Key words TNF-α, TLR2, ELISA, Immunoblotting, FACS, Quantitative real-time RT-PCR

1 Introduction

Mycobacterium tuberculosis, one of the world's leading causes of mortality and morbidity, has been attributed to foster a complex cytokine milieu during the infection [1, 2]. Roles of these myriad cytokines, be it the robust pro-inflammatory or the dampening anti-inflammatory cytokines, during the mycobacterial pathogenesis have been studied and evaluated. Tumor necrosis factor-alpha (TNF-α) is one such pro-inflammatory cytokine that is crucial to determine the outcome of the mycobacterial infection. TNF-α is not only requisite for the formation and maintenance of granuloma but also to execute the antimicrobial functions exhibited by the antigen presenting cells such as the macrophages [3, 4]. Mice deficient in TNF-α or TNF-α receptor (TNFR) display delayed granuloma formation and possess diminished ability to contain mycobacterial infection [5, 6]. Further, TNF-α and TNFR signaling also regulate the T cell functions [7]. Interestingly, mycobacteria-induced TNF-α modulates

Jagadeesh Bayry (ed.), *The TNF Superfamily: Methods and Protocols*, Methods in Molecular Biology, vol. 1155, DOI 10.1007/978-1-4939-0669-7_12,

other signaling networks [8, 9] that are believed to resurface during infections to fine-tune the immune responses [10–12]. Sonic hedgehog (SHH) signaling and TLR2 signaling cascades represent two such TNF-α regulates pathways.

Canonical SHH signaling involves the binding of the ligand SHH to its cognate receptor Patched-1 (PTCH1). The ligand–receptor interaction alleviates the inhibition on a positive transducer of the pathway, Smoothened (SMO). As a consequence, there is inactivation of a cytosolic inhibitory complex composed of glycogen synthase kinase-3β (GSK-3β) and activation of GLI1 transcription factor which is otherwise repressed by NUMB. GLI1 induces the expression of SHH responsive genes [13, 14].

Mycobacteria-mediated TLR2 signaling is rather complex. TLR2 engagement with mycobacteria renders the host to induce several signaling cascades and immunological responses [15, 16]. Activation of phosphoinositide 3-kinase (PI3K)–Protein kinase C (PKC)–Mitogen-activated protein kinase (MAPK) on mycobacterial infection is well established to be TLR2 signaling dependent [17]. Further, TLR2 responsive downstream targets such as suppressor of cytokine signaling-3 (SOCS-3) [10], matrix metalloproteinase 9 (MMP-9) [18], and cyclooxygenase-2 (COX-2) [19] play significant roles in immunological control of mycobacterial infection.

During mycobacterial infection, TNF-α induces the activation of SHH signaling that feeds back to regulate TLR2 responses [20]. Thus, modulation and regulation of TNF-α production and TNF signaling could prove decisive for the pathogenesis of tuberculosis. Multiple methods exist to determine and analyze the quantitative and qualitative functions of the TNF-α produced during mycobacterial infection. The assessment of TNF-α in the serum of the patients and experimental culture supernatants by ELISA can be correlated to the pathophysiology of the disease. Interestingly, soluble TNF receptors (sTNFRs) can also be examined by ELISA as the anti-inflammatory sTNFRs neutralize the TNF bioactivity. Loss-of-function assays such as utilization of neutralizing antibodies, antagonists, RNA interference further support the analysis. Immunoblotting, FACS, and quantitative real-time RT-PCR are described as tools to evaluate the infection induced TNF-α, SHH, and TLR2 signaling. Together, this chapter provides a comprehensive understanding of techniques to evaluate and identify the status of TNF-α and its responsive signaling mediators during mycobacterial infections.

2 Materials

2.1 Cell Culture

1. Brewer thioglycollate (Himedia, India).
2. Mice: C57BL/6 strain.
3. 2 mL syringes, 21-gauge needle.

4. Macrophages from mouse peritoneal exudates.
5. Flat bottom 6-well cell culture plates.
6. Complete Dulbecco's Modified Eagle Medium (DMEM) (Life Technologies, USA): DMEM with 10 % fetal bovine serum (FBS).
7. L929 cells-conditioned medium.
8. 5 % CO_2 incubator to be maintained at 37 °C.

2.2 ELISA

1. 96-well flat bottom microtiter plates (Nunc-Immuno™ MicroWell™, Thermo Fisher Scientific, USA).
2. Matched anti-TNF-α antibody pairs (capture and biotinylated detection antibodies) and TNF-α Standard (BioLegend, USA).
3. 1× Phosphate-buffered saline (PBS): 8.0 g NaCl, 1.16 g Na_2HPO_4, 0.2 g KH_2PO_4, 0.2 g KCl in 1,000 mL water, Adjust pH to 7.4 with HCl, autoclave or filter sterilize.
4. Blocking Buffer: 1 % BSA in PBS or 1× Assay Diluent A in the kit (BioLegend).
5. Dilution Buffer: 0.05 % Tween-20, 0.1 % BSA in PBS, or 1× Assay Diluent A in the kit.
6. Wash Buffer: 0.05 % Tween-20 in PBS.
7. Streptavidin-conjugated to horseradish peroxidase (HRP) (Jackson ImmunoResearch, USA).
8. 1× 3, 3′, 5, 5′-tetramethylbenzidine (TMB) (Bangalore Genei, India).
9. Stop solution: 7 % H_2SO_4 in water.
10. Microplate reader capable of measuring absorbance at 450 and 570 nm.
11. Multichannel pipettes and sterile tips.

2.3 Loss-of-Function Experiments

1. Recombinant mTNF-α and hTNF-α (Peprotech, USA).
2. Anti-TNF-α antibody (SouthernBiotech, USA).
3. Control IgG antibody.
4. TNF-α antagonist (Calbiochem, Millipore, Germany).
5. Dimethyl sulfoxide solution (DMSO) (Sigma-Aldrich, USA).
6. 31-Gauge needle insulin syringes.
7. Betulinic acid (Calbiochem, Millipore, Germany).
8. siGENOME™ SMARTpool *Shh* and non-targeting (NT) siRNAs (Dharmacon, USA).
9. Oligofectamine (Invitrogen, USA).

2.4 Immunoblotting

1. Radio-Immunoprecipitation Assay (RIPA) lysis buffer: 50 mM Tris–HCl, pH 7.4, 1 % NP-40, 0.25 % sodium-deoxycholate, 150 mM NaCl, 1 mM EDTA, 1 mM PMSF, 1 μg/mL aprotinin, leupeptin, pepstatin each, 1 mM Na_3VO_4, 1 mM NaF.
2. Buffer A: 10 mM HEPES–KOH, pH 7.9, 10 mM KCl, 0.1 mM EDTA, 0.1 mM EGTA, 1 mM DTT, 0.5 mM PMSF.
3. Buffer C: 20 mM HEPES–KOH, pH 7.9, 0.4 M NaCl, 1 mM EDTA, 1 mM EGTA, 1 mM DTT, 1 mM PMSF.
4. Bradford reagent.
5. Sodium dodecyl sulfate-polyacrylamide gel electrophoresis (SDS-PAGE) gel components.
   (a) 12 % Resolving Gel: 9.9 mL water, 12 mL 30 % Acrylamide, 7.5 mL 1.5 M Tris–HCl, pH 8.8, 0.3 mL 10 % SDS, 0.3 mL 10 % Ammonium persulfate (APS), 0.012 mL Tetramethylethylenediamine (TEMED).
   (b) 5 % Stacking Gel: 6.8 mL water, 1.7 mL 30 % Acrylamide, 1.25 mL 1 M Tris–HCl, pH 6.8, 0.1 mL 10 % SDS, 0.1 mL 10 % APS, 0.01 mL TEMED.
6. Tris-Glycine Buffer (For 1,000 ml): Tris-3 g, glycine-18 g, SDS-1 g.
7. Transfer Buffer (For 200 ml): Tris-1 g, glycine-0.58 g, SDS-0.074 g, methanol-40 mL.
8. 1× TBST buffer: 20 mM Tris–HCl, pH 7.4, 137 mM NaCl, 0.1 % Tween 20.
9. Prestained SDS-PAGE Standards (Bio-Rad, USA).
10. Polyvinylidene fluoride (PVDF) membrane (Millipore, USA).
11. 5 % skimmed milk in 1× TBST and 5 % BSA in 1× TBST.
12. Matched specific primary and secondary antibodies [SHH, GLI1, PTCH1, NUMB, pGSK-3β, pp85, p4EBP-1, pPKCδ, pPKCα, pERK1/2, pp38, SOCS-3, MMP-9 (Cell Signaling Technologies, USA), COX-2, PCNA (Calbiochem, Germany), β-ACTIN (Sigma-Aldrich, USA), secondary antibodies conjugated to HRP (Jackson ImmunoResearch, USA)].
13. Enhanced Chemiluminescence substrate (Perkin Elmer, USA).

2.5 FACS Analysis

1. FACS buffer: 1× PBS, 0.5–1 % BSA, or 5–10 % FCS containing 0.1 % NaN_3/Sodium azide.
2. Fc block (eBioscience, USA): 1:50 dilution in FACS buffer.
3. Fixation buffer: 1–4 % paraformaldehyde solution.
4. Fixation/permeabilization Concentrate (eBioscience, USA).
5. Fixation/permeabilization Diluent (eBioscience, USA).

6. Permeabilization buffer (10×) (eBioscience, USA).
7. FACS tubes: Polystyrene round-bottom tube, 12×75 mm (BD Biosciences, USA).
8. Flow cytometer.

2.6 RNA Isolation

1. TRI Reagent® (Sigma-Aldrich, USA).
2. Chloroform.
3. Isopropanol.
4. 70 % Ethanol.
5. Diethyl pyrocarbonate (DEPC)-treated water: Dissolve 0.01 % DEPC (Sigma-Aldrich, USA) in water with overnight stirring and autoclave.
6. 20× 3-(*N*-morpholino) propanesulfonic acid (MOPS) Buffer (1,000 mL): 16.8 g MOPS–NaOH, pH 7, 1.62 g NaOAc, 8 mL 0.5 M EDTA, autoclave at low pressure and store in amber bottle.
7. RNA Gel (For 100 mL): Prepare 1.5 % agarose in 80 ml DEPC-treated water. Cool the solution and add 18 mL formaldehyde and 2.5 mL of 1× MOPS.
8. Sample loading buffer (1 mL): 350 μL formaldehyde, 100 μL formamide, 25 μL 20× MOPS, 0.2 μL ethidium bromide, 525 μL DEPC-treated water.

2.7 First-Strand cDNA Synthesis

1. Oligo$(dT)_{18}$ primer.
2. cDNA synthesis kit (Bioline, UK): 5× buffer, RNase inhibitor, 200 U/μL M-MuLV Reverse Transcriptase.
3. Primer-RNA mix: 2 μg RNA, 1 μL oligo(dT)18 primer and make up the volume to 13 μL with DEPC-treated water
4. RT mix: 4 μL 5× buffer, 2 μL 10 mM dNTPs, 1 μL RNase inhibitor, and 0.1 μL of 200 U/μL M-MuLV Reverse Transcriptase.
5. DEPC-treated water.
6. Thermal cycler.

2.8 Quantitative Real-Time RT-PCR

1. ABI 7900HT real-time machine.
2. KAPA™ SYBR® FAST qPCR kit (KAPA Biosystems, USA): 2× qPCR master mix, ROX reference dye High.
3. Gene-specific forward and reverse primers (*see* Table 1).
4. Reaction mix: 4 μL of 2× qPCR master mix, 0.2 μL ROX reference dye High (compatible to ABI 7900HT machine), 1 μL of 5 nM gene-specific forward and reverse primers.

Table 1
List of primers used for quantitative real-time RT-PCR analysis

Sr. No.	Gene name	Sequence
1	*Gapdh* forward *Gapdh* reverse	5′-GAGCCAAACGGGTCATCATCT-3′ 5′-GAGGGGCCATCCACAGTCTT-3′
2	*Shh* forward *Shh* reverse	5′-AAAGCTGACCCCTTTAGCCTA-3′ 5′-TTCGGAGTTTCTTGTGATCTTCC-3′
3	*Gli1* forward *Gli1* reverse	5′-CCAAGCCAACTTTATGTCAGGG-3′ 5′-AGCCCGCTTCTTTGTTAATTTGA-3′
4	*Gli2* forward *Gli2* reverse	5′-CAACGCCTACTCTCCCAGAC-3′ 5′-GAGCCTTGATGTACTGTACCAC-3′

3 Methods

3.1 Isolation of Primary Macrophages from Peritoneal Exudates of Mice

1. Inject 1 mL of 8 % Brewer thioglycollate intraperitoneally (i.p.) into mice.
2. After 4 days of injection, euthanize the mice and collect the cells from lavage of peritoneal cavity with ice-cold 1× PBS using 21-gauge needle syringe into a 50 mL centrifuge tube.
3. Pellet the cells at 300 × *g*, 4 °C for 5 min and resuspend in complete DMEM medium.
4. Enumerate the number of cells using a hemocytometer and seed in complete DMEM according to requirement.

3.2 Isolation of Bone Marrow-Derived Macrophages (BMDMs)

1. Collect the bone marrow from femurs and tibias of mice using 21-gauge needle syringe.
2. Culture the cells in DMEM medium containing 30 % of L929 cell-conditioned medium for 7 days.
3. Use adherent cells as BMDMs. Confirm the purity of these cells by F4/80 staining using FACS (*see* Subheading 3.6.4).

3.3 Maintenance of Macrophage Cell Line

1. Culture murine RAW 264.7 macrophage-like cells or human monocytic THP1 cells in complete DMEM.
2. Differentiate human monocytic THP1 cells to macrophages by treatment with 5 nM of PMA for 18 h and rest the cells for 3 days prior to infection or treatment.
3. Seed desired number of macrophages/well in a 6-well plate according to the experiment to be set up.

3.4 Serum/Cell Supernatant Collection

1. Collect the blood samples from patients or healthy donors in a serum-separating tube or a vacutainer and allow it to stand for 30 min at room temperature to clot. Centrifuge at 1,000–2,000 × *g*

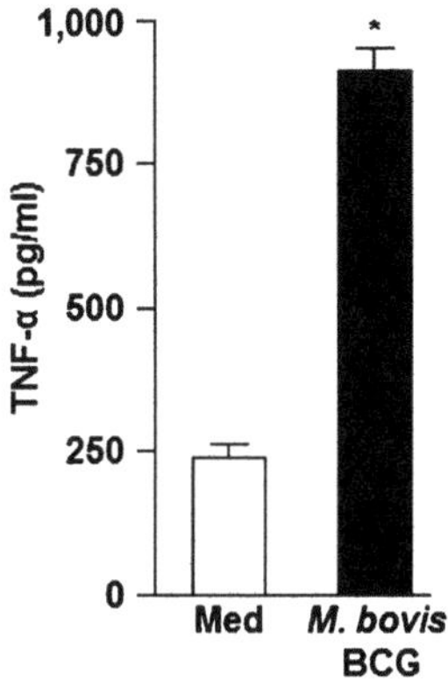

Fig. 1 Mycobacterial infection induces TNF-α secretion. Macrophages were infected with *M. bovis* BCG for 12 h and total secreted TNF-α protein levels was measured in cell-free supernatants by ELISA. Data represented were obtained from three independent experiments. *, $P < 0.05$ versus uninfected macrophages. *Med* medium [Copyright © 2013, American Society for Microbiology. Mol. Cell. Biol., 33, 2013, 543–556, DOI: 10.1128/MCB.01108-12]

for 10 min to remove the clot and store the obtained serum immediately at −80 °C after snap freezing.

2. Collect the experiment supernatants and spin at 2,500 × *g* for 10 min at 4 °C to remove the debris and store below −20 °C.

3.5 Estimation of TNF-α by ELISA

The extent of TNF-α or sTNFR produced during mycobacterial infection are evaluated by standard ELISA procedure. Patient sera and culture supernatants from in vitro/ex vivo/in vivo experiments are used as samples. In case of in vitro/ex vivo experiments, peritoneal macrophages or BMDMs or cell lines are infected with 10:1 MOI of mycobacteria for 12 h and supernatants are collected (Fig. 1a) (*see* **Note 1**).

1. Dilute the capture antibody provided in the dilution buffer or 1× coating buffer to a concentration of 1 μg/mL (or dilution suggested by the manufacturer to be used).
2. Coat the wells of the microtiter plate with 100 μL of the prepared capture antibody and leave it overnight at 4 °C (*see* **Note 2**).
3. Aspirate and wash the coated wells 3–4 times thoroughly with around 200 μL of wash buffer. After the final wash, remove the residual buffer by tapping the microtiter plate upside down on to the paper towel (*see* **Note 3**).
4. Add 300 μL of blocking buffer to each of the wells and leave at room temperature for 1 h on a plate shaker.
5. Meanwhile, keep the samples and standard ready. Prepare serial dilutions of the TNF-α standard starting from 500 pg/mL in the dilution buffer. Use dilution buffer alone as 0 pg/mL.

6. Aspirate and wash the wells as indicated before. Add 100 μL of the samples and standard to the respective wells and incubate at room temperature for 2 h (*see* **Note 4**).
7. Aspirate and wash the wells as indicated before. Dilute the detection antibody with dilution buffer or 1× Assay Diluent A to a concentration of 0.5 μg/mL (or dilution suggested by the manufacturer to be used). Add 100 μL of detection antibody to each well and incubate at room temperature for 2 h (*see* **Note 5**).
8. Aspirate and wash the wells as indicated before. Prepare 1:15,000 Streptavidin–HRP in dilution buffer. Add 100 μL of the antibody to each of the wells; incubate at room temperature in dark for 30 min.
9. Aspirate and wash the wells 4–5 times thoroughly. Add 100 μL of 1× TMB to each well and incubate in the dark till the samples turn blue (*see* **Note 6**).
10. Stop the reaction by adding 100 μL of 7 % H_2SO_4 into each well and gently mix the solution. The solution turns yellow.
11. Read the absorbance or OD of the reactions at 450 and 570 nm. Use $Absorbance_{450}$–$Absorbance_{570}$ ($\Delta OD_{450-570}$) as the value for calculations.
12. Calculation: Perform the calculation for the values obtained using MS Excel or manually. Subtract the $\Delta OD_{450-570}$ of the blank (Dilution buffer alone) with each of the values to negate the background. Plot the standard graph with known concentrations at *X*-axis (x) and respective $\Delta OD_{450-570}$ at *Y*-axis (y). Determine the "m" value for the reaction $y = mx$ by calculating the slope. Using the obtained "m" and Sample $\Delta OD_{450-570}$, calculate the concentration of each of the test samples.

3.6 Assessment of TNF-α Signaling and TLR2-Dependent Responses

Loss-of-function assays aid to identify TNF-α/TLR2 signaling mediators and target genes, which are examined using the classical immunoblotting, FACS, and quantitative real-time RT-PCR techniques. Immunoblotting and FACS involve utilization of antibodies to determine the qualitative and quantitative levels of the proteins, especially the target genes. Quantitative real-time RT-PCR provides the mRNA profile of the same.

3.6.1 TNF-α-Mediated SHH Signaling

TNF-α-induced signaling events can be examined after time and dose titration analysis. 100 ng/mL of TNF-α treatment for 8 h significantly induces SHH signaling in mouse peritoneal macrophages (Fig. 2a–c). Involvement of TNF-α can be further confirmed using antibody or antagonist-mediated blockade of the TNF-α and TNF-α receptor through both in vitro/in vivo experiments.

1. Treat macrophages with neutralizing antibody to TNF-α (10 μg/mL) or control IgG and TNF-α antagonist (5 μg/mL) or DMSO for 6 h prior to mycobacterial infection (MOI 10:1)

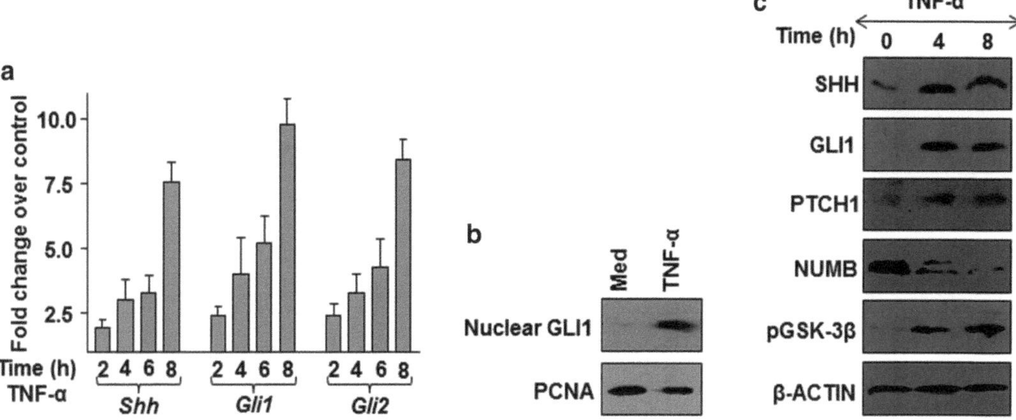

Fig. 2 TNF-α induces SHH signaling. (**a**). Peritoneal macrophages were treated with TNF-α at indicated time points and expression of *Shh*, *Gli1*, and *Gli2* transcripts were examined by quantitative real-time RT-PCR analysis (mean ± SE, $n=3$). (**b**). Nuclear fraction from TNF-α-treated macrophages was analyzed for GLI1 nuclear translocation by immunoblotting. (**c**). Macrophages were treated as described in (**a**) and immunoblotting was performed on total cell lysate with antibodies against SHH, GLI1, PTCH1, NUMB, and pGSK-3β. All blots are representative of three independent experiments. *Med* medium [Copyright © 2013, American Society for Microbiology. Mol. Cell. Biol., 33, 2013, 543–556, DOI: 10.1128/MCB.01108-12]

for 12 h. The samples can be used for immunoblotting/FACS or quantitative real-time PCR (Fig. 3a–c).

2. For in vivo experiment, inject 0.1 mg of anti-TNF-α antibody or control IgG per mice or 0.5 mg of TNF-α antagonist or DMSO per mice by i.p. 12 h prior to mycobacterial (10^8 bacteria) i.p. injection for 12 h. Isolate the peritoneal macrophages to analyze the TNF-α responses (Fig. 3d–f).

3.6.2 Restriction of TLR2 Signaling by TNF-α-Induced SHH

Pharmacological and siRNA-mediated knock down of SHH signaling can be carried out to confirm the role of TNF-α and SHH signaling in abrogating the mycobacteria-induced TLR2 responses. Pharmacological inhibitors are small molecule inhibitors that target the signaling mediators, whereas siRNA-mediated knock down involves transient transfection of the specific siRNA to silence the gene of interest.

1. Pharmacological inhibition.
   (a) Seed 5×10^6 macrophages per well in a 6-well plate in antibiotic-free complete DMEM.
   (b) Pretreat the macrophages with 10 μM of betulinic acid (pharmacological inhibitor of GLI1, a transducer of SHH signaling) or DMSO for 1 h. Add 100 ng of TNF-α for 8 h followed by mycobacteria infection (MOI 10:1) for 12 h.
   (c) Harvest the macrophage for TLR2 signaling assessment by immunoblotting/FACS techniques (Fig. 4a, b).

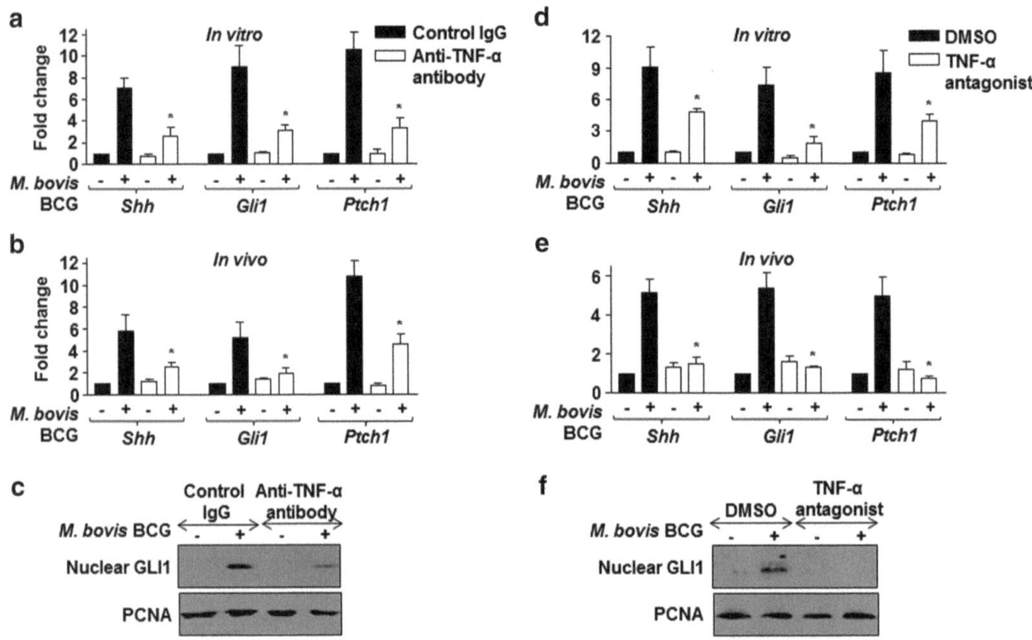

Fig. 3 TNF-α is a mediator of mycobacteria-trigged SHH signaling. (**a** and **b**). Peritoneal macrophages were treated in vitro (**a**) or mice were i.p. injected (**b**) with anti-TNF-α antibody prior to *M. bovis* BCG infection/injection to examine SHH signaling components by quantitative real-time RT-PCR (mean ± SE, $n=3$) *, $P<0.05$ versus control IgG treated/injected *M. bovis* BCG-infected macrophages. (**c**). Activation status of SHH signaling was ascertained by nuclear translocation of GLI1 with or without in vitro treatment of anti-TNF-α antibody. Blots are representative of three independent experiments. (**d–f**). Quantitative real-time RT-PCR and immunoblotting to analyze *M. bovis* BCG-induced SHH signaling induction after TNF-α antagonist pretreatment in experimental setup similar to (**a–c**). *Med* medium [Copyright © 2013, American Society for Microbiology. Mol. Cell. Biol., 33, 2013, 543–556, DOI: 10.1128/MCB.01108-12]

2. Transient transfection of macrophages.
   (a) Seed 0.5×10^6 RAW 264.7 or THP-1 macrophages in 6-well plates in complete DMEM. Meanwhile, make a premix of 100 nM of *Shh* or NT siRNA and oligofectamine (1 μL/reaction) in serum-free DMEM and incubate for 30 min.
   (b) Aspirate the complete DMEM from the 6-well plates, wash with 1× PBS and then add the premix with siRNA, swirl to mix, and place in the incubator for 8 h. Aspirate the siRNA mix and incubate the cells with complete media for 48–72 h.
   (c) Replace the complete media with antibiotic free complete DMEM and treat the cells with 100 ng of TNF-α for 8 h followed by mycobacterial infection (MOI 10:1) for 12 h. Harvest the cells and analyze the TLR2 responses by immunoblotting/FACS (Fig. 4c).

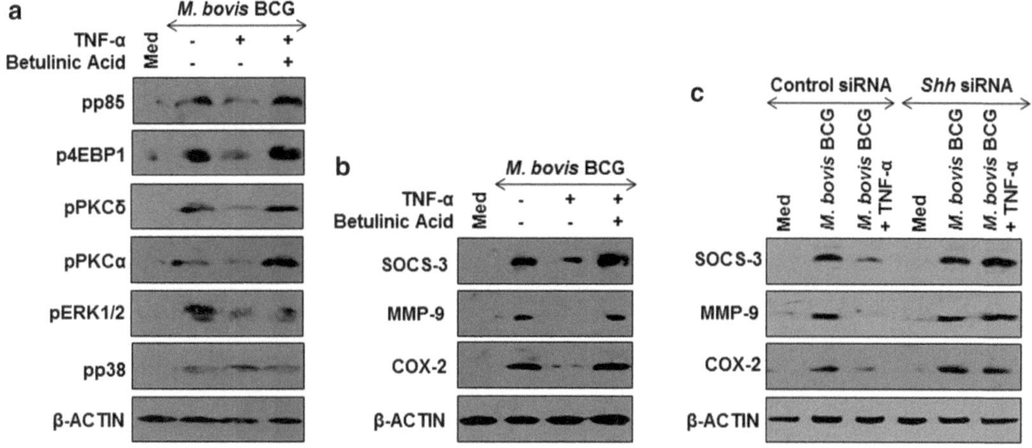

Fig. 4 TNF-α regulates TLR2-dependent responses. (**a** and **b**). Betulinic acid treated macrophages were stimulated with TNF-α followed by infection with *M. bovis* BCG. Immunoblotting was performed using specific antibodies to TLR2 signaling dependent (**a**) phosphorylated forms of p85, 4EBP1, PKCδ, PKCα, ERK1/2, p38 and (**b**) expression of SOCS-3, MMP-9, and COX-2. (**c**). *Shh* siRNA-transfected RAW 264.7 cells were treated with TNF-α followed by infection with *M. bovis* BCG to analyze indicated TLR2-dependent genes by immunoblotting. Blots are representative of three independent experiments. *Med* medium, *NT* non-targeting [Copyright © 2013, American Society for Microbiology. Mol. Cell. Biol., 33, 2013, 543–556, DOI: 10.1128/MCB.01108-12]

3.6.3 Immunoblotting Analysis

1. Preparation of total cell lysate (*see* **Note 7**).
   (a) Collect the culture supernatant and wash the macrophages with 1 mL of ice-cold 1× PBS. Scrap the cells in 1 ml of ice-cold 1× PBS and transfer the cells into 1.5 microfuge tubes and centrifuge at 2,500 ×*g* for 10 min at 4 °C.
   (b) Resuspend the pellet in 100 μL of RIPA buffer and incubate on ice for 30 min (*see* **Note 8**). Collect the whole cell lysate (supernatant) in a fresh microfuge tube after centrifuging at 2,500 ×*g* for 10 min at 4 °C.
   (c) Estimate the total protein in the lysate using Bradford assay.
2. Preparation of cellular and nuclear fraction.
   (a) Resuspend the cell pellet in 150 μL of Buffer A, gently mix and incubate on ice for 15 min. Add 7 μL of 10 % NP-40, vortex the sample for 30 s to 1 min till the solution turns clear.
   (b) Collect the cytosolic fraction (supernatant) in a fresh microfuge tube after centrifuging at 2,500 ×*g* for 15 min at 4 °C.
   (c) Lyse the pellet in 50 μL of Buffer C and nuclear lysate is collected (supernatant) in a fresh microfuge tube after centrifuging at 2,500 ×*g* for 15 min at 4 °C. Estimate the cytosolic and nuclear protein lysate using Bradford assay.

3. SDS-PAGE.
   (a) Prepare the SDS-PAGE gel and Tris-Glycine buffer to run the gel (*see* **Note 9**).
   (b) Run equal amount of the protein (total cell lysate or cytosolic/nuclear fractions) on SDS-PAGE. Load the prestained SDS-PAGE standards for reference.
4. Semi-dry transfer.
   (a) Cut required region of the gel (based on the molecular weight of the protein) with pre-stained marker as reference.
   (b) Cut the PVDF membrane according to the size of the cut gel, prime the membrane with methanol. Equilibrate the gel, blotting pads, and membrane in transfer buffer for few minutes (*see* **Note 10**).
   (c) Prepare a "sandwich" on the western blot semi-dry transfer apparatus in the following sequence blotting pad, equilibrated membrane, equilibrated gel, second piece of the blotting pad. Carry out the transfer at 200 mA for 4 h (*see* **Note 11**).
5. Developing blots.
   (a) After transfer, place the blots in 5 % of skimmed milk prepared in 1× TBST for 1 h to block the nonspecific binding sites. After washing with 1× TBST buffer, add the recommended concentration of primary antibody made in 5 % of BSA to the blots and incubate overnight at 4 °C (*see* **Note 12**).
   (b) Further, wash with 1× TBST buffer, incubate with suitable matched secondary antibody conjugated to HRP for 2 h (*see* **Note 13**).
   (c) Wash the membrane with 1× TBST and develop the blot using Enhanced Chemiluminescence system. Use β-ACTIN as loading control, whereas PCNA as nuclear fraction loading control (*see* **Note 14**).

3.6.4 Staining Surface Expressed Molecules for FACS Analysis

1. Pellet the cells at 300 × *g* for 5 min at 4 °C, wash and resuspend in ice-cold FACS buffer to obtain 1–5 × 10^6 cells/mL concentration of cells (*see* **Note 15**).
2. Collect cells in the FACS tubes (*see* **Note 16**).
3. Add 100 μL of Fc block to each sample and incubate on ice or at 4 °C for 20 min. Centrifuge at 300 × *g* for 5 min at 4 °C and discard supernatant.
4. Add recommended concentration (0.1–10 μg/mL) of the primary labeled antibody for direct staining. In case of

optimization of conjugated dyes required for staining the cells, make appropriate dilution of the antibody in FACS buffer. Incubate for at least 30 min at room temperature or 4 °C in the dark (*see* **Note 17**).

5. Centrifuge at 300×*g* for 5 min, wash with ice-cold FACS buffer thrice, and resuspend in 200 μL to 1 mL of ice-cold FACS buffer.
6. If the primary antibody used for staining was unlabeled, resuspend the cells in fluorochrome-conjugated secondary antibody containing FACS buffer, and incubate for 20–30 min at room temperature or 4 °C in the dark.
7. Wash the cells in ice-cold FACS buffer three times by centrifuging at 300×*g* for 5 min and resuspend in 200 μL to 1 mL of ice-cold FACS buffer.
8. Add 100 μL of fixation buffer and incubate for 10–15 min at room temperature to fix the cells (*see* **Note 18**).
9. Wash the cells in ice-cold FACS buffer by centrifuging at 300×*g* for 5 min and resuspend in 200 μL to 1 mL of ice-cold FACS buffer.
10. Place the cells in the dark on ice or at 4 °C till they are acquired for analysis.
11. Acquire the samples in a flow cytometer. Use specific controls to set the PMT voltage and compensation. Unstained cells, isotype controls, or blocking controls are necessary to acquire the data.
12. Based on fluorescent intensities, generate a series of subsets called gates to analyze the desired population of cells during data acquisition [21] (*see* **Note 19**).

3.6.5 Staining of Intracellular Molecules for FACS Analysis

1. Harvest the cells, wash in ice-cold FACS buffer, and pellet the cells at 300×*g* for 5 min at 4 °C.
2. Resuspend the cells in 1 mL of Fixation/Permeabilization buffer, vortex briefly (not more than 30 s), and incubate in dark for 60 min.
3. Centrifuge at 300×*g* for 5 min at 4 °C, resuspend pellet in 2 mL of 2× permeabilization buffer, and vortex briefly.
4. Centrifuge and wash again with 2× permeabilization buffer.
5. Add 100 μL of Fc block in 2× permeabilization buffer to each sample and incubate on ice or at 4 °C for 20 min.
6. Without washing the cells, add recommended concentration (0.1–10 μg/mL) of the primary antibody made in 1× permeabilization buffer and incubate at 4 °C for 30 min in dark. Centrifuge and wash cells in 1× permeabilization buffer.

7. If the primary antibody used for staining was unlabeled, resuspend the cells in fluorochrome-conjugated secondary antibody containing 1× permeabilization buffer and incubate for 20–30 min at room temperature or 4 °C in the dark.
8. Centrifuge and decant the supernatant. Leave about 100 μL of permeabilization buffer in the tube along with the stained cells for acquiring through FACS.
9. Acquire the samples in a flow cytometer as described in **steps 11** and **12** in Subheading 3.6.4.

3.6.6 RNA Isolation for Quantitative Real-Time RT-PCR (See Note 20)

1. Collect the culture supernatant and wash the macrophages with 1 mL of ice-cold 1× PBS.
2. Lyse the cells directly in 6-well plate using 600 μL of TRI reagent. Pipette the lysate up and down several times carefully and collect in a 1.5 mL microfuge tube.
3. Add 120 μL of chloroform to the lysate and vortex the samples for 1 min vigorously. Allow the samples to stand at room temperature for 10 min and centrifuge the samples at 2,500 × *g*, 4 °C for 20 min for phase separation (*see* **Note 21**).
4. Carefully collect the aqueous layer into a fresh microfuge tube and add 0.6 volume of isopropanol (200 μL), mix gently, and incubate at room temperature for not more than 45 min.
5. Pellet the precipitated RNA by centrifuging at 2,500 × *g*, 4 °C for 20 min.
6. Wash the pellet with 100 μL of 70 % ethanol at 2,500 × *g*, 4 °C for 10 min.
7. Air-dry the pellet and dissolve in 25 μL of DEPC-treated water.
8. Estimate the RNA quantity using the spectrophotometer (NanoDrop) and check for the quality and integrity of the RNA isolated by running the samples on an agarose gel.

3.6.7 First-Strand cDNA Synthesis

1. Use 1–2 μg of RNA for first-strand cDNA synthesis in a 20 μL reaction.
2. Make an initial primer-RNA mix in 0.2 ml microfuge tubes. Use a thermal cycler to heat the samples at 95 °C for 5 min (*see* **Note 22**).
3. Immediately chill the samples on ice for 5 min and then add the RT mix to each of the tubes with the initial primer-RNA mix.
4. Place the samples in a thermal cycler with the following program: 37 °C for 90 min, 70 °C for 10 min, 4 °C for ∞.
5. Dilute the samples to a total volume of 200 μL using sterile water.

3.6.8 Quantitative Real-Time RT-PCR

1. Use 384-well/96-well real-time plates for setting the reactions based on the compatibility of the real-time instrument. Add 2 μL of the diluted cDNA into wells.
2. Master reaction mix is prepared separately for each gene and added to the respective wells. Place the sealant carefully on the plate and centrifuge the plate at 600 × *g* at room temperature for 10 min.
3. Run samples on the real-time machine using the following program: 95 °C for 15 min for initial denaturation, 40 cycles of 95 °C for 15 s, annealing temperature for 15 s, 72 °C for 30 s and finally single cycle of 95 °C for 15 s, annealing temperature for 15 s, 72 °C for 15 s. Glyceraldehyde 3-phosphate dehydrogenase (GAPDH) or β-actin can be used as internal control (*see* **Note 23**).
4. Analyze the data using SDS 2.3 software (SABiosciences). Export the cycle threshold (Ct) values displayed against each of the samples in SDS 2.3 to Microsoft Excel for calculations. The differences in the Ct (ΔCt) values are calculated for each sample for a given gene by subtracting the Ct of treated vs control. ΔΔCt stands for relative normalization or difference in the ΔCts of each sample for a given gene with respective ΔCt of internal control like GAPDH for the same sample. The classical $2^{-\Delta\Delta CT}$ Method is used for obtaining the fold change [22].
5. Plot the values and perform statistical analysis using GraphPad Prism 5.0 (GraphPad software, USA).

4 Notes

1. Use all the reagents preferably after they have reached the room temperature.
2. Calculate the number of wells to be coated with capture antibody based on the samples. Each sample should be preferably analyzed in duplicates or triplicates.
3. Do not let the wells dry completely post aspiration, washing and tapping steps.
4. Standardize the optimum amount of the test samples to be used. If saturated values of the OD are obtained, dilute the samples in dilution buffer uniformly before use. Seal the plates, if necessary, before incubation at each step to avoid contamination or spillage.
5. Wash the wells thoroughly enough number of times to prevent high background reading or high variations.
6. Make sure that TMB addition is performed in dark to avoid nonenzyme-mediated substrate catalysis. Minimize the pipetting errors by using multichannel pipettes. Also, this prevents

variations in the reading, especially during the TMB step. TMB is carcinogenic. Handle with care.

7. Perform all the steps of sample preparation at 4 °C or on ice to maintain integrity of the proteins.
8. Make the RIPA lysis buffer mix freshly every time as the protease and phosphatase inhibitors have short shelf life. Store the buffer at 4 °C.
9. Make sure that Tris base is used and not Tris–HCl for preparing tris-glycine buffer or transfer buffer.
10. Priming the PVDF membrane with methanol for 5 min is an absolute requirement for making the PVDF membrane hydrophilic. Do not let the PVDF membrane to dry during any incubation processes.
11. Remove any air bubbles trapped between the gel and the membrane by rolling a glass rod over it.
12. Standardization of primary and secondary antibody concentrations to be used is necessary. Initially use the recommended concentration by the manufacturer.
13. Wash the blots thoroughly at the indicated steps to prevent background signals.
14. Prepare fresh mix of oxidase and luminol components of ECL to obtain strong signals (as recommended by the manufacturer).
15. Check the viability of the cells before the staining process. Around 90–95 % viability is recommended.
16. 2 ml Microfuge tube can be used instead of FACS tubes. However, do not use conical bottomed microfuge tubes to avoid concentration of cells at one place.
17. Do not add sodium azide to the buffers if the cells need to be recovered for functional assays as it can inhibit metabolic activities of the cell. Staining with ice-cold reagents/solutions and incubations at 4 °C give better results, since low temperature and presence of sodium azide prevents the modulation and internalization of surface antigens which can otherwise lead to a loss of fluorescence intensity and lower counts during acquiring the samples.
18. Make sure the pH of 1–4 % paraformaldehyde used for fixation is 7.4. This step is important as fixing of cells will stabilize the light scatter.
19. Carry out the FACS analysis on the same day as staining for better results. Analyze the change in forward and side scatter by gating and setting PMT voltage using appropriate controls and keep it constant for the rest of the tubes. Tubes with unstained cells are required to normalize the autofluorescence.

20. Wear powder-free gloves during RNA isolation procedure to prevent RNase from the skin contaminating the samples. Use RNase-free reagents, tips, microcentrifuges for the procedure.
21. Carefully take only the aqueous layer after the TRI reagent-chloroform spin for RNA precipitation as the interface contains the genomic DNA.
22. Use random hexamers for cDNA synthesis instead of oligo$(dT)_{18}$ primer if the integrity of the RNA was found not so good on the agarose gel. Add an additional step of 25 °C for 10 min to the thermal cycler program after initial denaturation.
23. Each sample needs to be set in duplicates for real-time analysis to rule out the pipetting error. Use gloves while placing the sealant on the real-time plate to avoid finger prints/impressions that can hinder the signal acquisition.

References

1. Cooper AM, Mayer-Barber KD, Sher A (2011) Role of innate cytokines in mycobacterial infection. Mucosal Immunol 4:252–260
2. Flynn JL, Chan J (2001) Immunology of tuberculosis. Annu Rev Immunol 19:93–129
3. Algood HM, Lin PL, Flynn JL (2005) Tumor necrosis factor and chemokine interactions in the formation and maintenance of granulomas in tuberculosis. Clin Infect Dis 41(Suppl 3):S189–S193
4. Keane J, Remold HG, Kornfeld H (2000) Virulent Mycobacterium tuberculosis strains evade apoptosis of infected alveolar macrophages. J Immunol 164:2016–2020
5. Bean AG, Roach DR, Briscoe H, France MP, Korner H, Sedgwick JD, Britton WJ (1999) Structural deficiencies in granuloma formation in TNF gene-targeted mice underlie the heightened susceptibility to aerosol Mycobacterium tuberculosis infection, which is not compensated for by lymphotoxin. J Immunol 162:3504–3511
6. Benini J, Ehlers EM, Ehlers S (1999) Different types of pulmonary granuloma necrosis in immunocompetent vs. TNFRp55-gene-deficient mice aerogenically infected with highly virulent Mycobacterium avium. J Pathol 189:127–137
7. Ehlers S, Kutsch S, Ehlers EM, Benini J, Pfeffer K (2000) Lethal granuloma disintegration in mycobacteria-infected TNFRp55-/- mice is dependent on T cells and IL-12. J Immunol 165:483–492
8. Oguma K, Oshima H, Aoki M, Uchio R, Naka K, Nakamura S, Hirao A, Saya H, Taketo MM, Oshima M (2008) Activated macrophages promote Wnt signalling through tumour necrosis factor-alpha in gastric tumour cells. EMBO J 27:1671–1681
9. Ohishi K, Varnum-Finney B, Serda RE, Anasetti C, Bernstein ID (2001) The Notch ligand, Delta-1, inhibits the differentiation of monocytes into macrophages but permits their differentiation into dendritic cells. Blood 98:1402–1407
10. Narayana Y, Balaji KN (2008) NOTCH1 up-regulation and signaling involved in *Mycobacterium bovis* BCG-induced SOCS3 expression in macrophages. J Biol Chem 283:12501–12511
11. Blumenthal A, Ehlers S, Lauber J, Buer J, Lange C, Goldmann T, Heine H, Brandt E, Reiling N (2006) The Wingless homolog WNT5A and its receptor Frizzled-5 regulate inflammatory responses of human mononuclear cells induced by microbial stimulation. Blood 108:965–973
12. Bansal K, Trinath J, Chakravortty D, Patil SA, Balaji KN (2011) Pathogen-specific TLR2 protein activation programs macrophages to induce Wnt-beta-catenin signaling. J Biol Chem 286:37032–37044
13. Ingham PW, McMahon AP (2001) Hedgehog signaling in animal development: paradigms and principles. Genes Dev 15:3059–3087
14. Ingham PW, Nakano Y, Seger C (2011) Mechanisms and functions of Hedgehog signalling across the metazoa. Nat Rev Genet 12:393–406
15. Koul A, Herget T, Klebl B, Ullrich A (2004) Interplay between mycobacteria and host signalling pathways. Nat Rev Microbiol 2: 189–202

16. O'Garra A, Redford PS, McNab FW, Bloom CI, Wilkinson RJ, Berry MP (2013) The immune response in tuberculosis. Annu Rev Immunol 31:475–527
17. Ghorpade DS, Leyland R, Kurowska-Stolarska M, Patil SA, Balaji KN (2012) MicroRNA-155 is required for *Mycobacterium bovis* BCG-mediated apoptosis of macrophages. Mol Cell Biol 32:2239–2253
18. Kapoor N, Narayana Y, Patil SA, Balaji KN (2010) Nitric oxide is involved in *Mycobacterium bovis* bacillus Calmette-Guerin-activated Jagged1 and Notch1 signaling. J Immunol 184:3117–3126
19. Bansal K, Narayana Y, Patil SA, Balaji KN (2009) *M. bovis* BCG induced expression of COX-2 involves nitric oxide-dependent and -independent signaling pathways. J Leukoc Biol 85:804–816
20. Ghorpade DS, Holla S, Kaveri SV, Bayry J, Patil SA, Balaji KN (2013) Sonic hedgehog-dependent induction of microRNA 31 and microRNA 150 regulates *Mycobacterium bovis* BCG-driven toll-like receptor 2 signaling. Mol Cell Biol 33:543–556
21. Ormerod MG (2008) Flow cytometry—a basic introduction. Lulu, Raleigh
22. Livak KJ, Schmittgen TD (2001) Analysis of relative gene expression data using real-time quantitative PCR and the 2(-Delta Delta C(T)) method. Methods 25:402–408

Chapter 13

Experimental Applications of TNF-Reporter Mice with Far-Red Fluorescent Label

Yury V. Shebzukhov, Anna A. Kuchmiy, Andrei A. Kruglov, Frauke Zipp, Volker Siffrin, and Sergei A. Nedospasov

Abstract

This chapter provides protocols for in vitro and in vivo analysis of TNF-producing cells from a novel TNF reporter mouse. In these transgenic mice, genetic sequence encoding far-red reporter protein Katyushka (FRFPK) was placed under control of the same regulatory elements as TNF, thus providing the basis for detection, isolation, and visualization of TNF-producing cells.

Key words TNF, Reporter mice, Far-red reporter protein Katyushka, Flow cytometry, Confocal microscopy

1 Introduction

Reporter organisms provide useful tools for identification, visualization, and fate mapping of cells producing specific gene products. Fluorescent proteins may also allow intra-vital analysis of such cells using two-photon microscopy and other imaging techniques (reviewed in refs. 1, 2). Additionally, fluorescent reporters may facilitate isolation of rare cell subsets that produce the protein of interest by FACS (fluorescence-activated cell sorting). Here we describe applications of one particular reporter system to follow TNF expression.

The first TNF reporter mouse was generated in early 1990s [3]. It was based on a transgene carrying 3,963-base pair (bp) DNA fragment from the TNF/LT genomic locus in which case TNF coding sequences and introns were replaced by cDNA sequence encoding chloramphenicol acetyltransferase (CAT). Using this early reporter system a constitutive synthesis of TNF in thymus [3], renal-specific induction of TNF expression by Shiga toxin [4], and UV-induced expression of TNF in skin [5] were demonstrated.

Jagadeesh Bayry (ed.), *The TNF Superfamily: Methods and Protocols*, Methods in Molecular Biology, vol. 1155, DOI 10.1007/978-1-4939-0669-7_13, © Springer Science+Business Media New York 2014

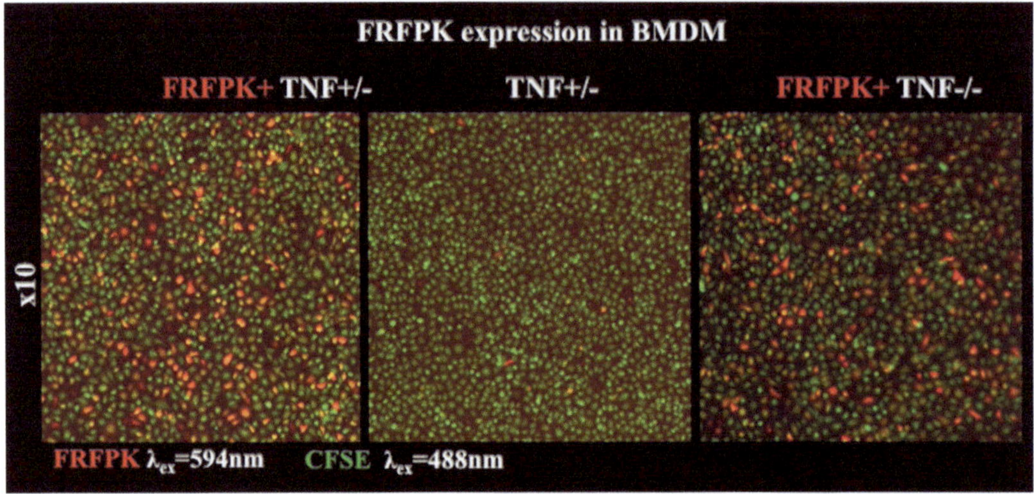

Fig. 1 FRFPK expression in BMDM detected by confocal microscopy. BMDM were stimulated for 9 h with 100 ng/mL of *E. coli* LPS. 3 h after beginning of LPS stimulation 5 μg/mL of BrefeldinA was added. Live (attached to surface) cells were visualized by confocal fluorescent microscope Zeiss LSM 710 at 10× magnification. Excitation of FRFPK was performed using 594 nm and detection by filter set for PE-Texas Red. CFSE was excited by 488 nm and detected by filter set for FITC

Another transgenic TNF reporter mouse was based on luciferase gene placed under control of TNF promoter [6], and this system was used for whole body imaging revealing TNF expression in response to stimulation by bacterial products (LPS and *N*-(3-oxo-dodecanoyl) homoserine lactone (C12)).

None of the above systems allowed intravital microscopy applications. We have recently described a novel TNF transgenic reporter mouse based on the far-red fluorescent protein Katyushka (FRFPK) [7]. Red fluorescence (emission maximum 635 nm) by this protein can be excited with yellow or yellow-green lasers in cytometry [7] and confocal microscopy setting (Fig. 1) or, as shown below, with >1,000 nm infrared frequencies in two-photon imaging settings (Fig. 2). In B6.FRFPK$^+$ reporter mouse, the gene encoding FRFP Katyushka was fused to the 3′ end of the murine TNF gene with retention of 3′-UTR required for correct posttranscriptional control [7]. Separation of TNF and FRFPK translation was achieved by utilization of the viral peptide system which rendered TNF produced from this transgenic insert inactive [7]. Below we provide protocols for in vitro and in vivo utilization of B6.FRFPK$^+$ reporter.

2 Materials

2.1 Equipments

1. Zeiss LSM 710 microscope.
2. Becton Dickinson LSRFortessa analyzer or FACS Aria II sorter equipped with a yellow-green laser.

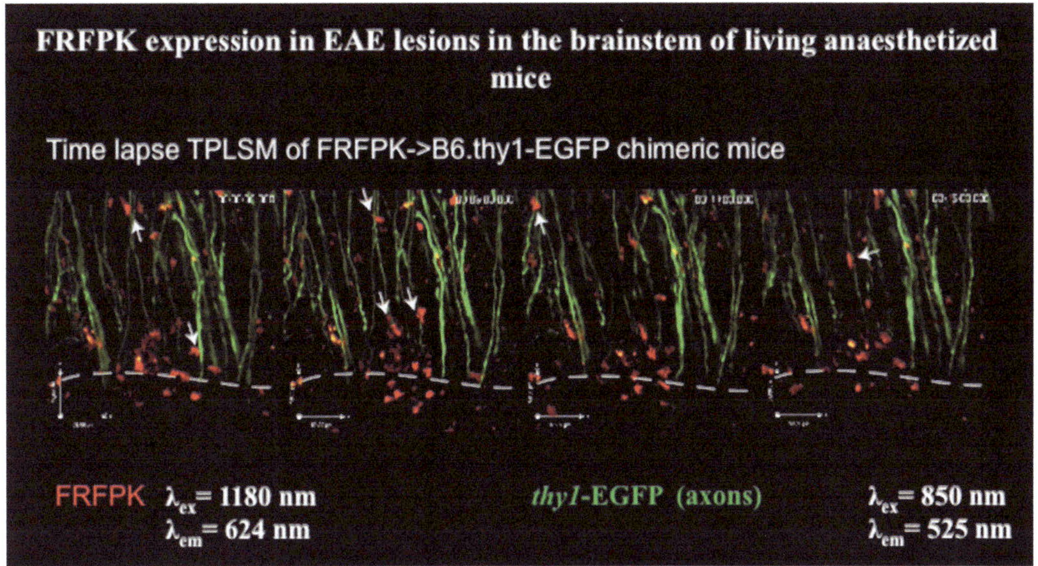

Fig. 2 Intravital TPLSM of a brainstem EAE lesion in a FRFPK+ → B6.thy1-EGFP mouse. EAE was induced and animals imaged at the peak of the disease. FRFPK+ lymphoid-like cells accumulate around vessel borders (*dashed line*) and invade the CNS tissue. Dynamic interactions of FRFPK+ (*arrow*) immune cells with axons (*green*) is visible in this time-lapse recording of 15 min (time in right upper corner, image acquisition every 60 s)

3. Dual NIR/IR Excitation Setup for Two-Photon Laser Scanning Microscopy (TPLSM) (TriMScope, LaVision BioTec, Germany).
4. MidiMACS and MiniMACS Separators (Miltenyi Biotec).
5. Ti:Sa laser (MaiTai, SpectraPhysics, Germany).
6. Synchronously pumped optical parametric oscillator (APE, Germany).
7. Harvard Apparatus Advanced Safety Respirator (Hugo Sachs, Germany).

2.2 Laboratory Animals, Reagents, Disposable Materials, and Solutions

1. B6.FRFPK+ transgenic mice (*see* **Note 1**), B6.*thy1*-EGFP mice, TNF-deficient mice. B6.FRFPK+ transgenic mice were on C57BL/6 background. For some experiments they were backcrossed to TNF-deficient background. For bone marrow chimeric mice, B6.*thy1*-EGFP mice on C57BL/6 background in which a *thy1* promoter drives the expression of enhanced green fluorescent protein (thus, axons/neurons express EGFP) were used as recipients. All animal experiments were performed in accordance with institutional, state, and federal guidelines. Mice were sacrificed by cervical dislocation.
2. Enrofloxacin (Baytril®, Bayer HealthCare).
3. Isoflurane (Abbott).
4. PBS–BSA–EDTA solution: PBS containing 0.1 % BSA, 2 mM EDTA.

5. Proteinase K (Biodeal, Markkleeberg, Germany).
6. Recombinant mouse macrophage colony-stimulating factor (M-CSF, Miltenyi Biotec). M-CSF was dissolved in PBS with 0.1 % BSA to stock concentration 10 μg/mL and aliquots were kept at −80 °C (*see* **Note 2**).
7. *E. coli* LPS (Sigma-Aldrich). LPS was dissolved in sterile water to stock concentration 1 mg/mL and aliquots were kept −20 °C (*see* **Note 2**).
8. Brefeldin A (Sigma-Aldrich). Brefeldin A was dissolved in ethanol to stock concentration 5 mg/mL and aliquots were kept −20 °C.
9. Complete DMEM (or RPMI 1640): DMEM (or RPMI 1640) with 10 % fetal calf serum (FCS), 100 U/mL penicillin, 100 μg/mL streptomycin, 2 mM L-glutamine.
10. Carboxyfluorescein diacetate, succinimidyl ester (CFSE). CFSE was dissolved in DMSO to stock concentration 5 mM and aliquots were kept at −80 °C (*see* **Note 2**).
11. Erythrocyte lysis buffer (Liter): 8.29 g NH_4Cl, 1 g $KHCO_3$, 200 μl 0.5 M EDTA, pH was adjusted with HCl to 7.2–7.4. Erythrocyte lysis buffer was sterilized by filtration through 0.2 μm filter.
12. 70 μm Cell strainer (BD Falcon).
13. 30 μm Pre-separation filter (Miltenyi Biotec).
14. 4-Well chamber slides (Thermo Scientific).
15. 10 cm Plastic Petri dishes.
16. LS or MS columns (Miltenyi Biotec).
17. Anti-FITC MultiSort Kit (Miltenyi Biotec).
18. Anti-PE, anti-Cy5/anti-Alexa Fluor 647 and anti-CD90.2 magnetic beads (Miltenyi Biotec).
19. Lymphocyte Separation Medium LSM 1077 (PAA).
20. MOG emulsion (per mouse): 100 μL water containing 150 μg MOG_{35-55} peptide (Pepceuticals, UK) emulsified in 100 μL complete Freund's adjuvant (CFA) supplemented with 400 μg of killed dessicated *M. tuberculosis* H37Ra (BD Difco).
21. Pertussis toxin (PTx, List Biological Laboratories).
22. Neutralizing monoclonal antibodies to IL-4 (clone 11B11), IL-12 (clone C17.8), IFNγ (AN 18) (eBioscience, San Diego, CA, USA).

3 Methods

3.1 Generation of Bone Marrow Chimeric Mice

1. For the generation of bone marrow chimeric mice, lethally irradiate the recipient animals with 1,100 cGy (split dose) [8].
2. Use B6.FRFPK$^+$ transgenic mice as donors. Sacrifice the donor animals; isolate the bone marrow cells by flushing of femur and tibia, and deplete CD90$^+$ T cells from single cell suspension by anti-CD90.2 MicroBeads according to the manufacturer's protocol. Briefly, count the cells and incubate with the appropriate amount of anti-CD90.2 beads: 10 μL of anti-CD90.2 MicroBeads and 90 μL of PBS–BSA–EDTA buffer per 10^7 total cells. Load the cells on the LS column. Wash the column additionally three times with 3 mL of PBS–BSA–EDTA buffer. Analyze the flow through (depleted fraction) bone marrow cells by flow cytometry for CD3 expressing cells, which is intended to be <1 % of lymphocyte gated cells.
3. Reconstitute the recipients with $12–20 \times 10^6$ T cells-depleted donor bone marrow cells about 8 h after irradiation. Keep the mice on 0.01 % Enrofloxacin in drinking water for about 4 weeks. Engraftment takes place over 8 weeks of recovery.

3.2 Experimental Autoimmune Encephalomyelitis (EAE)

For active EAE in B6.FRFPK$^+$ → B6.*thy1*.EGFP, immunize the mice subcutaneously with 150 μg of $MOG_{35–55}$ emulsified in CFA into the region drained by the inguinal lymph node. Inject 200 ng of PTx per mouse intraperitoneally at the time of immunization and 48 h later.

3.3 Preparation of Bone Marrow-Derived Macrophages (BMDM)

1. Flush bone marrow cells from femurs with cold DMEM. Centrifuge homogenized cell suspension for 5 min at 300 × *g* and resuspend in complete DMEM supplemented with 30 ng/mL of M-CSF (*see* **Note 3**).
2. Seed 6–7 ml of bone marrow cells (10^6 cells/ml) to the non-adhesive 10 cm plastic Petri dishes and incubate 7 days at 37 °C and 5 % CO_2.
3. Replace cultivation medium with cold Ca^{2+} and Mg^{2+} free PBS; incubate the cells for 20 min at 4 °C, wash them from the surface, collect and centrifuge cells for 5 min at 300 × *g*. Immediately use the collected cells for experiments or cultivate further (up to 3 days) in the same medium as above.

3.4 Preparation of CD4$^+$ T Cells

1. Isolate secondary lymphoid organs (spleens, and mesenteric, inguinal, and axillary lymph nodes) and smash them through the 70 μm cell strainer in PBS–BSA–EDTA solution, sediment cells at 300 × *g* at 4 °C for 10 min (*see* **Note 4**).
2. To lyse erythrocytes, resuspend cells in 2–5 mL of erythrocyte lysis buffer, incubate 5 min on ice and add 10× volume of

PBS–BSA–EDTA. Vigorously shake the cell suspension 5–6 times, filter through 70 μm cell strainer and sediment cells at 300 × *g* at 4 °C for 10 min.

3. Resuspend total lymphoid cell preparation either in complete RPMI 1640 medium for activation and analysis or isolate naive T cells for subsequent T cell polarization experiments.
4. To isolate naive CD4⁺ T cells, incubate total lymphocytes (2×10^8 cells/ml) for 15 min on ice with 10 μg/mL of antibodies against Fcγ receptor (to prevent unspecific binding), 2 μg/mL of FITC-conjugated anti-CD4 antibodies, 1 μg/mL of PE-conjugated anti-CD44 and Cy5-conjugated anti-CD25 antibodies.
5. Wash the cells in 10× volume of PBS–BSA–EDTA solution, resuspend at 2×10^8cells/mL with 10 % of anti-FITC MultiSort magnetic beads (part of Miltenyi Biotec anti-FITC MultiSort Kit) in PBS–BSA–EDTA solution and incubate for 15 min on ice.
6. Wash the cells with 10× volume of PBS–BSA–EDTA solution, resuspend at 2×10^8cells/mL, filter through 30 μm pre-separation filter and apply to LS or MS column depending on the total amount of cells (*see* **Note 5**).
7. Wash the cells three times with 3 mL (for LS column) or with 0.5 mL (for MS column) of PBS–BSA–EDTA solution, remove column from the magnet and elute positive fraction with 5 mL (for LS column) or 1 mL (for MS column) of PBS–BSA–EDTA solution.
8. Add 40 μl of Release reagent (part of Miltenyi Biotec anti-FITC MultiSort Kit) to 1 mL suspension of positive cells. Incubate the cells for 10 min on ice and then for 20 min at room temperature, sediment cells at 300 × *g* at 4 °C for 10 min and then resuspend 10^7 of cells with 60 μl of PBS–BSA–EDTA solution, 30 μl of Stop reagent (part of Miltenyi Biotec anti-FITC MultiSort Kit) and 5 μl of anti-PE and anti-Cy5/anti-Alexa Fluor 647 magnetic beads.
9. After 15 min incubation on ice, wash the cells in 10× volume of PBS–BSA–EDTA solution, resuspend at 2×10^8 cells/mL or minimum in 0.5 mL of PBS–BSA–EDTA solution, and apply to LS or MS column (*see* **Note 5**).
10. Wash the column as above and collect the negative cell fraction, containing naive $CD4^+CD25^-CD44^-$ cells.
11. For cell stimulation, immobilize anti-CD3 antibodies at 4 μg/mL concentration in PBS on the surface of cell culture-treated plasticware for at least 1 h at 37 °C. Wash the unbound anti-CD3 antibodies with PBS.
12. Sediment naive CD4⁺ T cells at 300 × *g* at 4 °C for 10 min and resuspend in complete RPMI 1640 medium with 20 μM of

2-mercapto-ethanol (*see* **Note 6**). Add anti-CD28 antibodies at concentration 1 μg/mL for co-stimulation of T cells.

13. To keep the cells in unpolarized (Th0) conditions, add anti-IL-4, anti-IL-12, and anti-IFNγ antibodies at final concentration of 10 μg/mL. Cytokine/antibody cocktails for polarization of T cells under various conditions can also be applied.
14. At day 2 after beginning of stimulations, transfer the cells into new culture plates and supplement the medium with 10 ng/mL of IL-2 (*see* **Note 7**). At day 6, collect the cells, carefully layer them on the LSM (*see* **Note 8**) and centrifuge 20 min at 2,000 × *g* (centrifuge breaks should be off) at room temperature. Collect live cells concentrated at interphase, wash with fresh culture medium, and use for flow cytometry analysis or sorting (*see* **Note 9**).

3.5 Microscopic Analysis of FRFPK+ BMDM

1. Incubate 5×10^6 BMDM for 10 min at room temperature in 1 mL of 5 μM CFSE solution in PBS containing 5 % FCS (PBS/FCS) and wash three times with 10 mL of PBS/FCS.
2. Resuspend BMDM in complete DMEM, seed in 4-well chamber slides, 7.5×10^5 cells in 1 mL per well, and incubate overnight at 37 °C and 5 % CO_2.
3. Next day, stimulate cells for 9 h with 100 ng/mL of *E. coli* LPS. 3 h after beginning of LPS stimulation, add 5 μg/mL of Brefeldin A.
4. Visualize live BMDM (attached to the bottom of slide chambers) by confocal fluorescent microscope Zeiss LSM 710 at 10× magnification. Perform excitation of FRFPK using 594 nm and detection by filter set for PE-Texas Red. Perform excitation of CFSE by 488 nm and detection by filter set for FITC (Fig. 1).

3.6 Anesthesia and Preparation of Imaging Field for Intravital Imaging

1. Anesthetize the mice using 1–1.5 % isoflurane in oxygen/nitrous oxide (2:1) with a facemask, then tracheotomize and continuously respirate with a Harvard Apparatus Advanced Safety Respirator.
2. Transfer the anesthetized animal to a custom-built microscopy table and fix in a hanging position. Expose the brainstem and superfuse with isotonic Ringer solution, which is continuously exchanged by a peristaltic pump.
3. Install a sterile agarose patch (0.5 % in 0.9 % NaCl solution) on the now-exposed brain surface to reduce heartbeat and breathing artifacts.

3.7 Intravital Two-Photon Laser Scanning Microscopy (TPLSM)

1. For intravital imaging use dual near infrared (NIR) (850 nm) and infrared (IR) (1,180 nm) excitation, i.e., simultaneous two-photon excitation of the sample with a Ti:Sa laser and a synchronously pumped optical parametric oscillator [9].

2. Collect XYZ stacks within a scan field of 300 × 300 μm at 512 × 512 pixel resolution and a *z*-plane distance of 2 μm at a frequency of 400 or 800 Hz. Applied laser powers should be ranged from 2 to 6 mW at the specimen's surface, which allows imaging depths up to 150 μm [9].
3. Perform detection of EGFP fluorescence at 525 nm and FRFPK fluorescence at 624 nm. FRFPK$^+$ cells (arrows, red) demonstrate even distribution of fluorescence within the cell and no signs of toxic cell damage (Fig. 2). They show lymphocyte morphology and motility in EAE lesions (*see* **Note 10**), exhibit interactions with axons in the brainstem EAE lesion and also accumulate around vessels (dashed line defines vessel border).

3.8 Stimulation of Cells for Flow Cytometry Analysis and Sorting

1. Stimulate BMDMs as follows. Replace cultivation medium in Petri dishes with primary macrophage culture with DMEM and stimulate the cells with 100 ng/mL of *E. coli* LPS in the presence of 5 μg/ml Brefeldin A. Alternatively, resuspend the collected BMDM in DMEM and seed at desired density 1 h before stimulation with LPS. 4–6 h after stimulation with LPS, replace the medium with cold PBS and collect the BMDM as described above.
2. Stimulate the T cells as follows. Collect the cells as in 3.4.14, centrifuge, resuspend in fresh complete RPMI 1640 medium with 20 μM of 2-mercapto-ethanol and treat for 6 h with 1 μg/mL of Ionomycin and 10 ng/mL of PMA in the presence of 5 μg/ml Brefeldin A.

3.9 Flow Cytometry Analysis of FRFPK$^+$ Cells

Perform the analysis of BMDM and T cells using BD LSRFortessa analyzer, equipped with yellow-green laser (excitation 561 nm) and filter sets for PE-Texas Red (band pass 600 LP, filter 610/20 nm) or PE-Cy5 (band pass 635 LP, filter 670/30 nm) (Figs. 3 and 4).

3.10 Sorting of FRFPK$^+$ Cells

1. For sorting, resuspend the cells in PBS–BSA–EDTA solution and pass through 30 μm pre-separation filter.
2. Perform sorting of FRFPK positive and negative BMDM and T cells using BD FACS Aria II sorter equipped with yellow-green laser and filter sets for PE-Texas Red or PE-Cy5. Typically, 15–30 % of stimulated cells are FRFPK$^+$ (Figs. 3 and 4) (*see* **Note 11**).
3. Use sorted cell fractions for transcriptional (qRT-PCR) and protein analysis (Western blot).

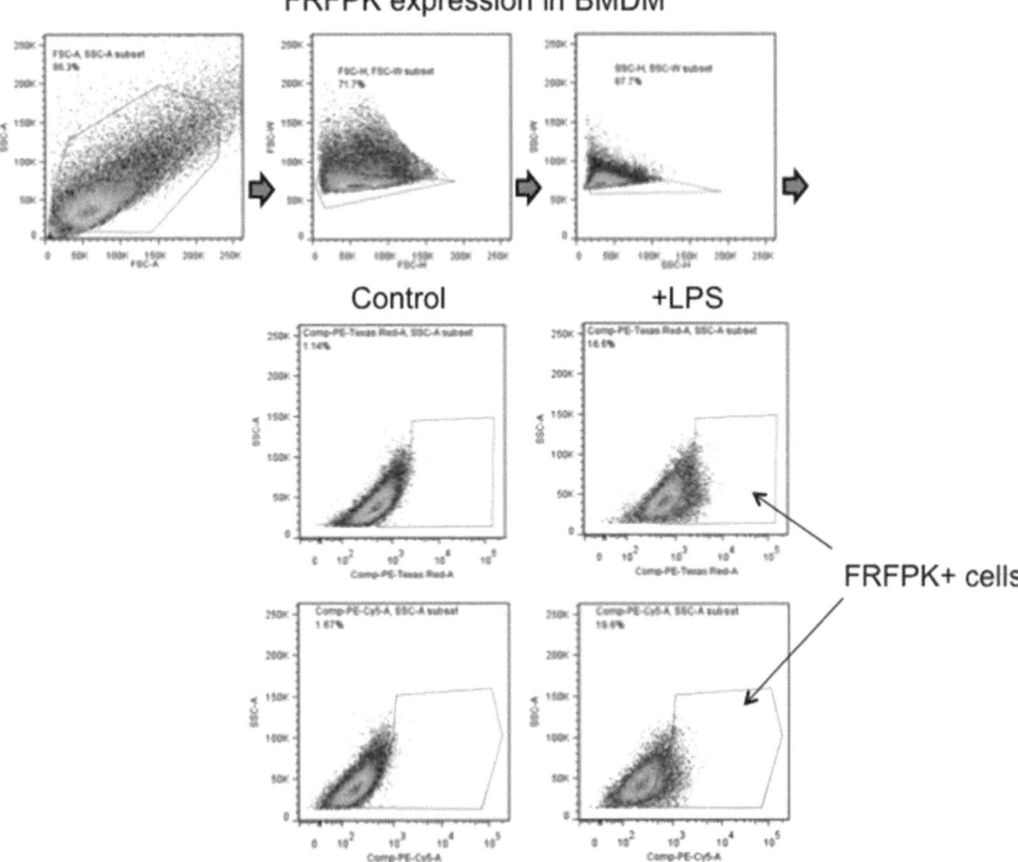

Fig. 3 FRFPK expression in BMDM detected by flow cytometry. BMDM were stimulated 4 h with 100 ng/mL of *E. coli* LPS in the presence of 5 μg/mL of BrefeldinA. FRFPK was exited by yellow–green laser at 561 nm. Cells were gated according Forward Scatter (FS) and Side Scatter (SS) parameters and FRFPK was detected at Texas Red (band pass 600 LP, filter 610/20 nm) and PE-Cy5 (band pass 635 LP, filter 670/30 nm) channels

4 Notes

1. B6.FRFPK+ mice were bred as heterozygotes. To genotype the mice, tail biopsies (5 mm) were cut and placed into the 1.5 mL tubes, filled with 500 μL of DNA isolation buffer (100 mM NaCl, 10 mM Tris–HCl pH 8, 25 mM EDTA, 0.5 % SDS), supplemented with 10 μL of 20 mg/mL Proteinase K and incubated at 56 °C with shaking until tissue is completely dissolved. Then 300 μL of 3 M NaCl were added, samples were vortexed and centrifuged 15 min at 20,000 × *g* and 4 °C. Supernatants were transferred to new tubes and mixed with 750 μL of isopropanol. DNA was precipitated by 30 min centrifugation at 20,000 × *g* and 4 °C, washed with 70 % ethanol, dried, resuspended in 400 μL of 5 mM Tris–HCl pH 8.5, and analyzed by PCR.

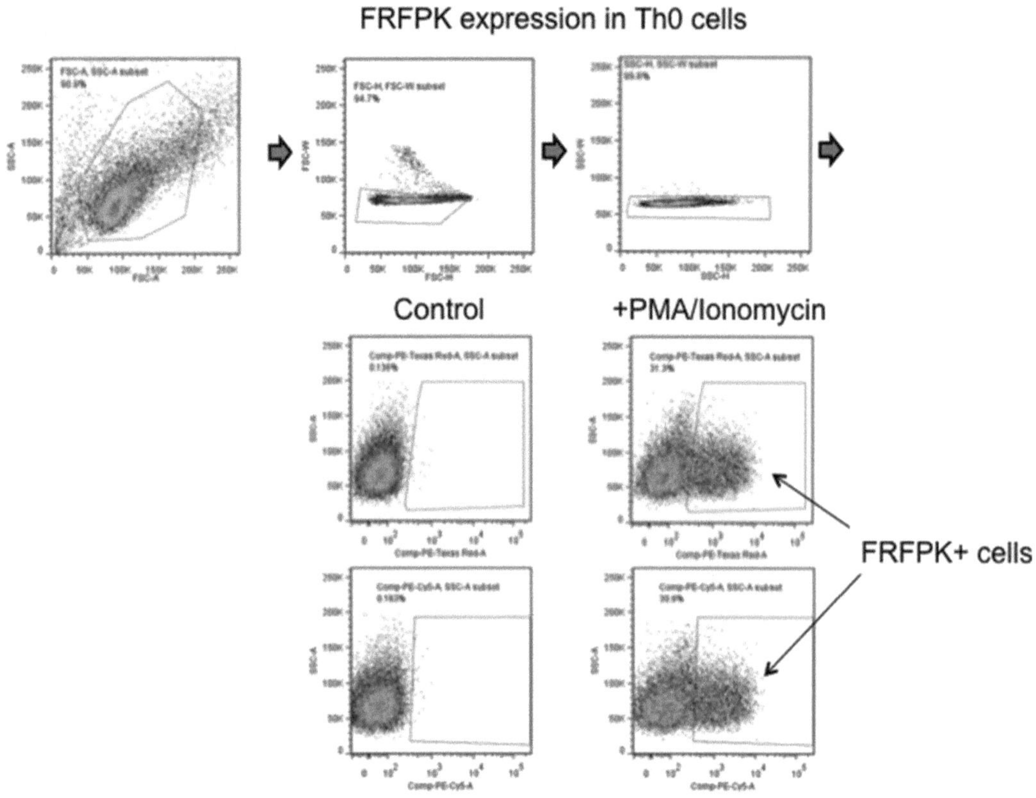

Fig. 4 FRFPK expression in T cells detected by flow cytometry. Th0 cells were stimulated 6 h with 1 μg/mL of Ionomycin and 10 ng/mL of PMA in the presence of 5 μg/mL of BrefeldinA. FRFPK was exited by yellow–green laser at 561 nm. Cells were gated according Forward Scatter (FS) and Side Scatter (SS) parameters and FRFPK was detected at Texas Red (band pass 600 LP, filter 610/20 nm) and PE-Cy5 (band pass 635 LP, filter 670/30 nm) channels

For 10 μL PCR reactions the following components were mixed:

1 μL of 10× PCR buffer

0.25 μL of dNTP mix (10 mM each)

0.5 U of Taq polymerase

0.25 μL of each primer (20 μM)

1–2 μL of template DNA

Water to 10 μL

Following primers are used for PCR reaction

TNF_(Kat)_wt_F: GGAGTCCGGGCAGGTCTACT
TNF_(Kat)_wt_R: GGTGCATGAGAGGCCCACAG
Kat_(Kat)_R: GCCCCCGTCTTCGTATGTGG

Reaction conditions: 5 min at 94 °C, 35 cycles of 94 °C 1 min, 63 °C 1 min, 72 °C 1 min; 5 min at 72 °C, hold at 4 °C.

PCR products: WT band was 258 bp, FRFPK^{+} band was 448 bp.

2. Thawed aliquots should not be frozen again.
3. Alternatively, DMEM with penicillin/streptomycin and L-Glutamine can be supplemented with 30 % of L929-conditioned medium and 20 % of horse serum.
4. It is critical that all centrifugation steps are performed in refrigerated centrifuge.
5. Up to 2×10^8 of total cells/10^7 of positive cells can be applied to the MS column, up to 2×10^9 of total cells/10^8 of positive cells can be applied to the LS column.
6. Presence of 2-mercapto-ethanol in culture medium is crucial for growth of T cells.
7. Fresh medium can be added if cell culture medium became yellow. 2–3 times dilutions are recommended depending on the color of the medium in cell culture.
8. 3 ml of LSM in 15 mL tube and 10 mL of LSM in 50 ml tube.
9. We also transfected T cells from reporter mice with MSCV retroviruses (based on the pMSCV-IRES-GFP vector) to overexpress sets of T cell lineage-specific and TCR-activated transcription factors. With such double reporter system (FRFPK for TNF and GFP for transcription factors), one can study the role of various transcription factors in the regulation of TNF expression in distinct T cell subsets.
10. Mice were imaged 2–3 days after onset of EAE. Accurate temperature control (35–37 °C) of the animal is essential to ensure cell motility. Excitation of FRFPK in our in vivo setup in the brainstem was most effective at frequencies between 1,150 and 1,200 nm.
11. Although these novel B6.FRFPK$^+$ reporter mice are yet at early stages of the analysis, several useful applications have been already demonstrated. Visualization of cells and quantification of TNF expression in various in vitro systems seems feasible. Importantly, due to design, TNF produced from reporter transgenic insert is inactive and does not contribute to the overall TNF activity/physiology. The option of fluorescence-based cell sorting should be useful for understanding TNF functions in specific cellular subsets, in particular in T cells. Finally, FRFPK$^+$ cells are perfectly usable for in vivo imaging. Advantages to be mentioned are the non-overlapping emission with CFP/GFP/YFP and the differentiability to orange fluorophores, e.g., tdRFP. Thus, parallel imaging of four and even more fluorophores seems doable without apparent side effects, which remains the problem for some other red fluorescent proteins.

References

1. Niesner RA, Hauser AE (2011) Recent advances in dynamic intravital multi-photon microscopy. Cytometry A 79A:789–798
2. Kuchmiy AA, Efimov GA, Nedospasov SA (2012) Methods for in vivo molecular imaging. Biochemistry (Mosc) 77:1339–1353
3. Giroir BP, Brown T, Beutler B (1992) Constitutive synthesis of tumor necrosis factor in the thymus. Proc Natl Acad Sci U S A 89: 4864–4868
4. Harel Y, Silva M, Giroir B, Weinberg A, Cleary TB, Beutler B (1993) A reporter transgene indicates renal-specific induction of tumor necrosis factor (TNF) by shiga-like toxin. Possible involvement of TNF in hemolytic uremic syndrome. J Clin Invest 92:2110–2116
5. Bazzoni F, Kruys V, Shakhov A, Jongeneel CV, Beutler B (1994) Analysis of tumor necrosis factor promoter responses to ultraviolet light. J Clin Invest 93:56–62
6. Kravchenko VV, Kaufmann GF, Mathison JC, Scott DA, Katz AZ, Grauer DC et al (2008) Modulation of gene expression via disruption of NF-κB signaling by a bacterial small molecule. Science 321:259–263
7. Kuchmiy AA, Kruglov AA, Galimov AR et al (2011) New TNF reporter mouse to study TNF expression. Russ J Immunol 5:205–214
8. Siffrin V, Brandt AU, Radbruch H, Herz J, Boldakowa N, Leuenberger T et al (2009) Differential immune cell dynamics in the CNS cause CD4+ T cell compartmentalization. Brain 132:1247–1258
9. Herz J, Siffrin V, Hauser AE, Brandt AU, Leuenberger T, Radbruch H et al (2010) Expanding two-photon intravital microscopy to the infrared by means of optical parametric oscillator. Biophys J 98:715–723

Chapter 14

A Solid-Phase Assay for Studying Direct Binding of Progranulin to TNFR and Progranulin Antagonism of TNF/TNFR Interactions

Qingyun Tian, Shuai Zhao, and Chuanju Liu

Abstract

The discovery that TNF receptors (TNFR) serve as the binding receptors for progranulin (PGRN) reveals the significant role of PGRN in inflammatory and autoimmune diseases, including inflammatory arthritis. Herein we describe a simple, antibody-free analytical assay, i.e., a biotin-based solid-phase binding assay, to examine the direct interaction of PGRN/TNFR and the PGRN inhibition of TNF/TNFR interactions. Briefly, a 96-well high-binding microplate is first coated with the first protein (protein A), and after blocking, the coated microplate is incubated with the biotin-labeled second protein (protein B) in the absence or presence of the third protein (protein C). Finally the streptavidin conjugated with a detecting enzyme is added, followed by a signal measurement. Also discussed in this chapter are the advantages of the strategy, key elements to obtain reliable results, and discrepancies among various PGRN proteins in view of the binding activity with TNFR.

Key words TNF, TNFR, Progranulin, Solid-phase assay, Biotin-labeled protein

Abbreviations

BSA	Bovine serum albumin
CpG-ODNs	CpG oligonucleotides
CSF	Cerebral spinal fluid
EBNA	Epstein-Barr nuclear antigen
FGF	Fibroblast growth factor
FTLD	Frontotemporal lobar degeneration
GEP	Granulin–epithelin precursor
HEK	Human embryonic kidney
HRP	Horseradish peroxidase
PCDGF	GP88/PC-cell derived growth factor
PEPI	Proepithelin
PGRN	Progranulin
TLR9	Toll-like receptor

Jagadeesh Bayry (ed.), *The TNF Superfamily: Methods and Protocols*, Methods in Molecular Biology, vol. 1155, DOI 10.1007/978-1-4939-0669-7_14, © Springer Science+Business Media New York 2014

TMB	3,3′,5,5″-Tetramethylbenzidine
TNFR	Tumor necrosis factor receptor
TNF-α	Tumor necrosis factor-α

1 Introduction

Solid-phase binding assay is a widely used approach to study protein–protein interactions in vitro, especially antibody-based solid-phase assay, in which antigen-specific antibodies are used [1–3]. However, there may be some issues associated with the antibody-based assay, limiting the use of the assay. One common problem is the nonspecific binding of the antibodies, which can result in the high background. Another concern comes from the inconsistency of the antibodies employed, especially when polyclonal antibodies are used. The ideal conditions need to be established for each batch, including antibody concentration, incubation time, dilution buffer, and so on. In addition, some antibodies may not be commercially available. In view of these potential issues, establishing a simple and an antibody-free approach is of importance for routinely detecting direct protein–protein interactions in vitro. Biotin-based solid-phase assay is some sort of a simple and time-saving technique used for protein interaction analysis, and more importantly, it is antibody-free. Briefly, the first protein (protein A) is immobilized to an ELISA plate (solid phase). After blocking, biotin-labeled second protein (protein B), alone or together with the third protein (protein C), is added to the well. After washing, bound protein is detected by addition of a streptavidin conjugated with an enzyme followed by a detection step and signal measurement (Figs. 1, 2, and 3).

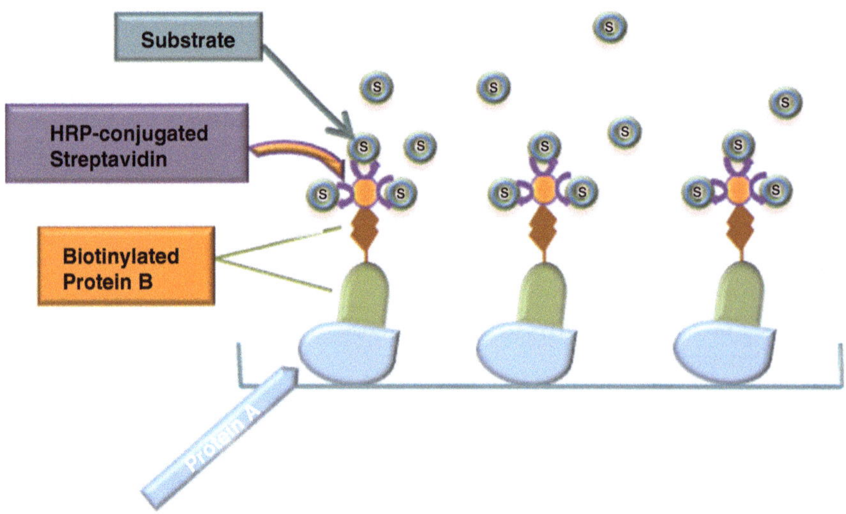

Fig. 1 Schematic diagram of biotin-based solid-phase binding assay

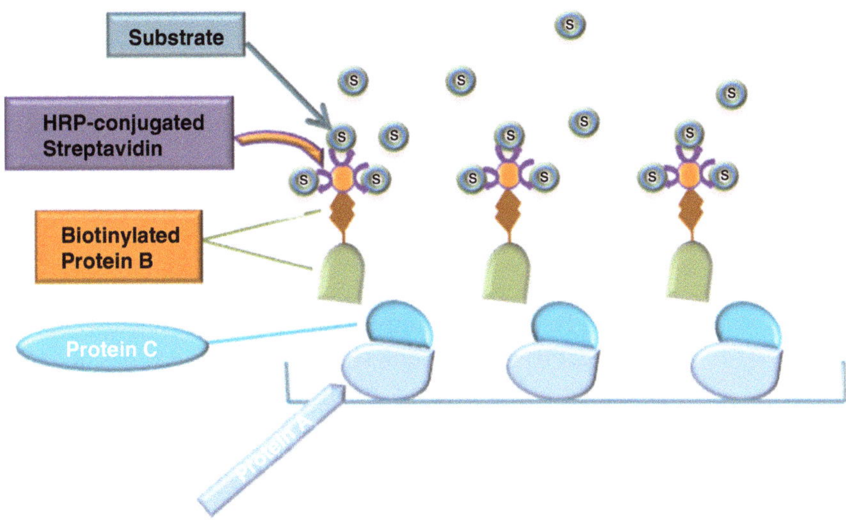

Fig. 2 Schematic diagram of biotin-based solid-phase inhibition assay

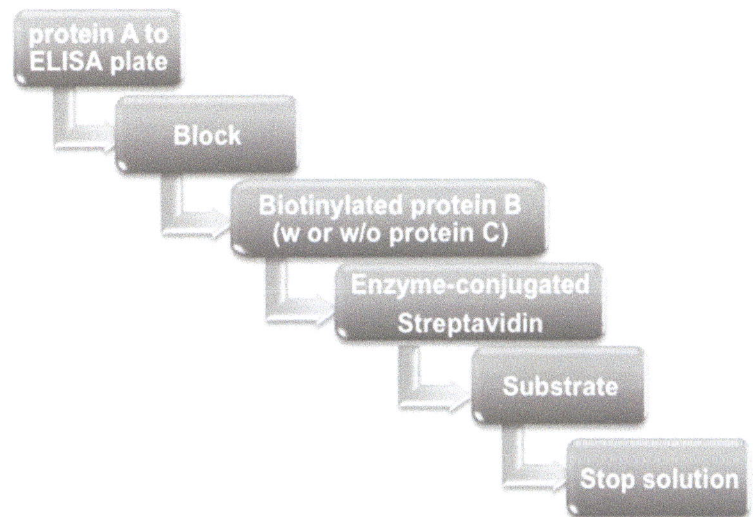

Fig. 3 Pipeline of detecting protein–protein interactions using biotin-based solid-phase assay

Progranulin (PGRN), also known as proepithelin (PEPI), PC cell-derived growth factor (PCDGF), and granulin epithelin precursor (GEP), is a 593-amino-acid autocrine growth factor [4, 5]. PGRN is highly expressed in epithelial cells, certain types of neurons, and macrophages [6], and is likewise expressed in a broad range of other tissues and cell types including skeletal muscle, chondrocytes, adipose tissue, hematopoietic cells, and immune cells, including T cells and dendritic cells [7–9]. PGRN has multiple physiological functions, and is involved in many types of disease processes, including autoimmune disorders, cancer, and neurodegenerative diseases. PGRN promotes cell proliferation, and is crucial to the development and generation of fast-growing epithelial, endothelial cells, and cancer cells [10]. PGRN levels are elevated in many types of cancer, including breast cancer, ovarian cancer, and cholangiocarcinoma [11–13]. Mutations of the *PGRN* gene are known to lead to the development of frontotemporal lobar degeneration (FTLD) [14, 15], and the protein levels of PGRN in serum and cerebral spinal fluid (CSF) were significantly reduced in both unaffected and affected PGRN-mutation carriers in FTLD [16–18]. The importance of PGRN has also been emerging in respect to the immune system [19]. PGRN has been reported to promote $CD4^+$ T cells to differentiate into $Foxp3^+$ regulatory T cells [20] and to enhance antibody diversity in B cells [21]. PGRN also binds with Toll-like receptor 9 (TLR9) and assists in the recruitment of CpG oligonucleotides (CpG-ODNs) in macrophages [22], a process which is critical for the elimination of infection.

Our recent studies have led to the isolation of PGRN as a binding partner of TNFR in a functional genetic screen [20, 23]. We have confirmed the interactions between PGRN and TNFR using various approaches, including Co-IP, FastStep Kinetic Assay, and antibody-based solid-phase assay [20]. Here we describe a biotin-based solid-phase binding assay using PGRN and TNFR as examples. In addition, a biotin-based assay for examining PGRN inhibition of TNF binding to TNFR is also presented.

2 Materials

1. Recombinant proteins.
   (a) Recombinant human progranulin: Purified from progranulin stable cell line-PGRN HEK293 EBNA stable cell line (Lab progranulin) [24], or R&D Systems Inc. (Mouse myeloma cell line NS0-derived, Thr18-Leu593).
   (b) Recombinant Human TNF RII (Invitrogen or R&D Systems Inc.), also known as CD120b and p75.
   (c) Recombinant Human TNF-α (R&D systems Inc.): *E. coli-derived*, Val77-Leu233.
2. EZ-Link Sulfo-NHS-Biotinylation Kit (Thermo scientific) (*see* **Note 1**).
3. Zeba™ Spin Desalting Column: It contains a high-performance resin that offers exceptional desalting or buffer exchange for protein samples.
4. TBS buffer: Tris-buffered saline, 10 mM Tris–HCl, pH 7.4, 150 mM NaCl.
5. Binding buffer: TBS buffer containing 0.1 % (w/v) BSA, 1 mM $CaCl_2$. *Prepare fresh.*

6. Streptavidin-HRP (Invitrogen): Diluted 1:2,500 in binding buffer.
7. TMB Solution (Invitrogen): Ready to use, no dilution or further preparation required. In the presence of HRP, TMB will turn to blue.
8. Stop solution: 2 N Sulfuric Acid.
9. Blocking solution: 5 % (w/v) bovine serum albumin (BSA) Fraction V, Heat Shock Treated (Fisher scientific), 0.05 % (w/v) sodium azide in TBS. Prepare sodium azide as 20 % stock solution, then add to TBS buffer, dissolve BSA to the buffer, mix gently up and down. *Prepare fresh.*

3 Methods

3.1 Biotinylation of Protein

Sulfo-NHS-Biotin is a popularly used water-soluble biotinylation reagent for labeling antibodies, proteins, and any other macromolecules that contain primary amine. It can be conjugated to many proteins of the primary amino groups to form stable amide bonds without perturbing the function of the molecule due to the small size of biotin, and it has high binding affinity to streptavidin and avidin.

The manufacturer's protocol is followed to label proteins. Briefly, calculate the amount of reagent needed, and prepare the biotin solution; add to the protein solution for labeling reaction, then perform buffer exchange to remove excess biotin reagent using a desalting column, followed by an HABA assay [2-(4-hydroxyazobenzene) benzoic acid] to estimate biotin label incorporation.

3.2 Purification of Recombinant PGRN Protein

1. Culture the progranulin stable cell line-PGRN HEK293 EBNA in tissue culture dishes in DMEM supplemented with 10 % of fetal bovine serum (FBS) and 50 μg/mL of G418 (Geneticin). After reaching confluence, add serum-free medium to the cells, and continue incubation for another 2 days, then collect the medium.
2. To purify recombinant PGRN protein, add nickel-chelating resin to the medium and mix at 4 °C for overnight. After washing, the bound protein was eluted and analyzed by SDS-PAGE and Western blot with anti-PGRN antibody.

3.3 Solid-Phase Binding Procedure for Detecting the Binding of PGRN to TNFR

1. Coat various doses (0–0.5 μM) of PGRN (protein A) or BSA (serving as control) in TBS buffer containing 0.5 % BSA to ELISA plate (100 μL/well) (*see* **Note 2**). Perform the assay using triplicates for each set of samples (*see* **Note 3**). Cover the plate and store at room temperature overnight. The plate can be stored at 4 °C for several days.

2. Discard the solution in the well and add 100 μL of blocking solution to each well using a multichannel pipette. After soaking for 1 min, invert the plate over a sink and strike it onto adsorbent paper toweling to remove the solution.
3. Add 200 μL of blocking solution to each well using a multichannel pipette. Leave at room temperature for 1–3 h (*see* **Note 4**).
4. Discard the blocking solution, and wash the wells three times with binding buffer.
5. Flick out the final wash. Invert the plate and tap it hard onto adsorbent paper towel to remove residual liquid (*see* **Note 5**).
6. Dilute the biotin-labeled TNFR2 (protein B) in binding buffer to a concentration of 1 ng/μL. Add 100 μL of this solution to each experimental well. Cover the plate and incubate at 37 °C for 2–3 h.
7. Aspirate the wells to remove unbound protein. Wash the wells three times with binding buffer. Remove residual buffer as in **step 5**.
8. Dilute streptavidin–HRP reagent 1:2,500 in binding buffer. Add 100 μL of this reagent to each well using a multichannel pipette. Incubate the plate for 15 min at room temperature (*see* **Note 6**).
9. Aspirate the wells to remove unbound streptavidin–HRP. Wash the wells seven to ten times with binding buffer. Remove residual buffer as in **step 5**.
10. Add 100 μL of TMB to each well using a multichannel pipette. Keep the plate in the dark until a blue color is obtained (typically 5–30 min) (*see* **Note** 7).
11. Stop the reaction by adding 100 μL of 2 N H_2SO_4 solution to each well using a multichannel pipette.
12. Read the plate using an automatic plate reader at 450 nm (*see* **Notes 8–11**) and analyze the data (*see* Fig. 4a).

3.4 Solid-Phase Binding Procedure for Detecting the Inhibition of TNF/TNFR Interactions by PGRN

1. Coat ELISA plate with TNFR2 (100 μL/well) in TBS buffer containing 0.2 % BSA. The well without TNFR2 serves as a blank. Perform the assay using triplicates for each set of samples. Cover the plate with plastic film and store at room temperature overnight or 4 °C for up to 1 week.
2. Follow the **steps 2–5** in Subheading 3.3.
3. Add various doses (0–1 μM) of PGRN (protein C) or BSA (serving as control) in binding buffer to each experimental well. Cover the plate with plastic film and incubate at 37 °C for 1 h. Then add biotin-labeled TNFα (protein B) to each experimental well and incubate at 37 °C for another 2 h.
4. Follow the **steps 7–12** in Subheading 3.3 (Fig. 4b).

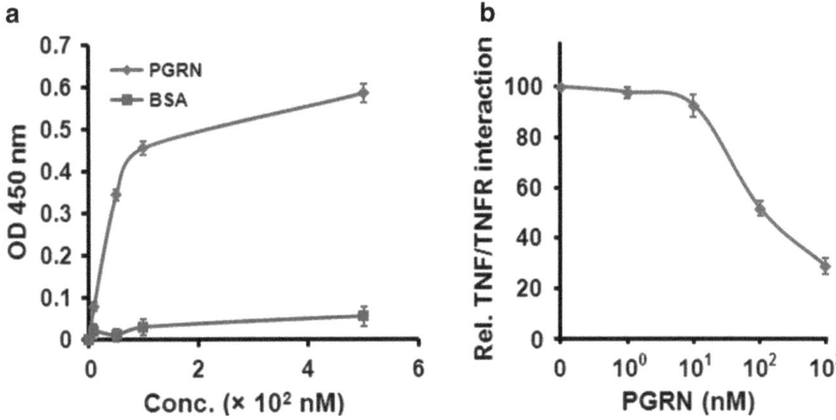

Fig. 4 Reprehensive examples of solid-phase assay. (**a**) Binding of PGRN to TNFR2. Various doses of PGRN or BSA, as indicated, were absorbed to an ELISA plate. After blocking biotin-labeled TNFR2 was added, and the bound protein from the liquid phase was then detected by streptavidin conjugated with horseradish peroxidase. (**b**) PGRN inhibition of TNFα binding to TNFR2. After blocking, plates precoated with TNFR2 were incubated with various amounts of PGRN, as indicated, and biotin-labeled TNFα, and the bound biotin-labeled TNFα was detected as described in (**a**)

4 Notes

1. EZ-Link Sulfo-NHS-Biotin is moisture-sensitive. Prepare the biotin reagent immediately before use.
2. The addition of BSA is not always required when preparing protein samples. And when it is used, the concentration in dilution buffer should be optimized from various trials aiming at the particular target protein and reagents. An appropriate selection of the buffer helps to block nonspecific binding sites to the upmost and retain the specific binding affinity at the samc timc.
3. Replicating samples for each set is preferred to get reliable results. Usually, the samples should be performed using at least triplicates, if possible, six to eight replicates, the more the better. The experiments are also expected to be repeated for several times, to ensure that the results are reproducible.
4. Blocking is the principal procedure to reduce nonspecific binding. A preferable blocking effect results from the proper choice of blocking reagents and the proper dosage of it. Different reagents may bring about dramatically varied background. The same blocker may perform distinctly different in different tests. Nonfat dry milk, for example, is regarded as an effective blocking buffer in certain kinds of solid-phase assay, like antibody-based assay, which, however, is not the same situation as in this biotin-based solid-phase assay. In this study

we used BSA as the blocker. To determine the optimal BSA concentration, preliminary experiments should be performed with different concentrations of BSA in blocking buffer. Meanwhile, a longer blocking time should be expected in case of high background binding.

5. Another basic way to get rid of high background is to present enough washes. Insufficient washing will lead to reagent residual in the wells, which may attenuate the difference between control and experimental groups. It is critical to guarantee plenty of washing times and complete elimination of the washing buffer each time.
6. To avoid excessive HRP signal, it is important to perform multiple dilutions of streptavidin-HRP before use. A feasible concentration of HRP can be figured out by performing experiments with different dilutions.
7. The signal development is time-dependent in the assay. After adding the substrate to start coloration, the plate should be kept in the dark for maintaining the stability of TMB. But it is important to check the plate every few minutes. Stop solution should be added whenever the coloration is obvious. Overtime incubation might result in overreactions. A typical time period is 5–30 min, which relies on different properties, concentrations, and binding affinities of reagents.
8. A highly reliable OD450 value is conventionally considered as ranging between 0.2 and 1.5. The values exceeding this range may be not reliable. High OD value may occur because of improper and inefficient washing or high concentration of reagents. Low OD value is probably due to insufficient incubation time, improper sample dilutions, expiration of reagents, or incorrect operation on ELISA readers (such as using wrong wavelength).
9. Another reason that leads to low OD value is probably due to the use of sodium azide. Sodium azide is an inorganic compound; it has an inhibitory effect on the activity of HRP and can quench peroxidase activity and suppress the growth of bacteria during incubations. It may be added for its bacteriostatic properties, but should be avoided by the step of incubation with HRP reagents.
10. In summary, biotin-based solid-phase assay is a simple, cost-effective analytical assay for studying protein interactions. For instance, it is antibody free and significantly reduces the workload, as there is no need to optimize antibody conditions. More importantly, it avoids the inconsistency and nonspecific binding resulted from the application of various antibodies. Although the biotin-based solid-phase assay offers numerous advantages over other assays, one potential disadvantage

associated with this assay is that the biotinylation process may alter the structure and properties of the proteins of interest, and the labeled protein may thus exhibit less or no binding activities to its binding partners.

11. The fact that PGRN is composed of 593 amino acids but displays an apparent molecular weight of approximately 90 kDa indicates that PGRN is modified, including glycosylation [25]. It appears that the various glycosylation and conformation are probably critical for the binding of PGRN to TNFR, since recombinant PGRN produced from a HEK-EBAN stable line, but not purchased from R&D System (data not shown), bound to TNFR (Fig. 4a). These findings are in line with the binding activities of perlecan, a PGRN-binding glycoprotein [26]. Perlecans purified from different cell types or same cell type with different expression systems have been shown to have the potential for significant variation in glycosylation and function, and can vary dramatically in FGF receptor-based assays or in growth factor binding assays [27–29]. It is worthwhile to note that the PGRN-binding activities of TNFR2 obtained from several commercial sources also significantly varied. For instance, TNFR2 purchased from Invitrogen or Sigma, but not from R&D System, showed reproducible binding with PGRN (data not shown), indicating that proper conformation of TNFR is also important for their ligand binding properties.

Acknowledgements

This work was supported partly by NIH research grants R01AR062207, R01AR061484, R56AI100901, and a Disease Targeted Research Grant from Rheumatology Research Foundation. Researches in the laboratory of C. J. Liu were also supported by Atreaon, Inc.

References

1. Aroca-Aguilar JD, Sanchez-Sanchez F, Ghosh S, Fernandez-Navarro A, Coca-Prados M, Escribano J (2011) Interaction of recombinant myocilin with the matricellular protein SPARC: functional implications. Invest Ophthalmol Vis Sci 52:179–189
2. Graham NA, Pope MD, Rimchala T, Huang BK, Asthagiri AR (2007) A microtiter assay for quantifying protein-protein interactions associated with cell-cell adhesion. J Biomol Screen 12:683–693
3. Arlinghaus FT, Eble JA (2012) The collagen-binding integrin alpha2beta1 is a novel interaction partner of the T. flavoviridis venom protein flavocetin-A. J Biol Chem 288:947–955
4. Liu CJ (2011) Progranulin: a promising therapeutic target for rheumatoid arthritis. FEBS Lett 585:3675–3680
5. Liu CJ, Bosch X (2012) Progranulin: a growth factor, a novel TNFR ligand and a drug target. Pharmacol Ther 133:124–132
6. Daniel R, He Z, Carmichael KP, Halper J, Bateman A (2000) Cellular localization of gene expression for progranulin. J Histochem Cytochem 48:999–1009

7. Wang D, Bai X, Tian Q, Lai Y, Lin EA, Shi Y et al (2012) GEP constitutes a negative feedback loop with MyoD and acts as a novel mediator in controlling skeletal muscle differentiation. Cell Mol Life Sci 69:1855–1873
8. Matsubara T, Mita A, Minami K, Hosooka T, Kitazawa S, Takahashi K et al (2012) PGRN is a key adipokine mediating high fat diet-induced insulin resistance and obesity through IL-6 in adipose tissue. Cell Metab 15:38–50
9. Elkabets M, Gifford AM, Scheel C, Nilsson B, Reinhardt F, Bray MA et al (2011) Human tumors instigate granulin-expressing hematopoietic cells that promote malignancy by activating stromal fibroblasts in mice. J Clin Invest 121:784–799
10. He Z, Bateman A (1999) Progranulin gene expression regulates epithelial cell growth and promotes tumor growth in vivo. Cancer Res 59:3222–3229
11. Frampton G, Invernizzi P, Bernuzzi F, Pae HY, Quinn M, Horvat D et al (2012) Interleukin-6-driven progranulin expression increases cholangiocarcinoma growth by an Akt-dependent mechanism. Gut 61:268–277
12. Serrero G (2003) Autocrine growth factor revisited: PC-cell-derived growth factor (progranulin), a critical player in breast cancer tumorigenesis. Biochem Biophys Res Commun 308:409–413
13. Diaz-Cueto L, Arechavaleta-Velasco F, Diaz-Arizaga A, Dominguez-Lopez P, Robles-Flores M (2012) PKC signaling is involved in the regulation of progranulin (acrogranin/PC-cell-derived growth factor/granulin-epithelin precursor) protein expression in human ovarian cancer cell lines. Int J Gynecol Cancer 22: 945–950
14. Baker M, Mackenzie IR, Pickering-Brown SM, Gass J, Rademakers R, Lindholm C et al (2006) Mutations in progranulin cause tau-negative frontotemporal dementia linked to chromosome 17. Nature 442:916–919
15. Cruts M, Gijselinck I, van der Zee J, Engelborghs S, Wils H, Pirici D et al (2006) Null mutations in progranulin cause ubiquitin-positive frontotemporal dementia linked to chromosome 17q21. Nature 442:920–924
16. Ghidoni R, Benussi L, Glionna M, Franzoni M, Binetti G (2008) Low plasma progranulin levels predict progranulin mutations in frontotemporal lobar degeneration. Neurology 71:1235–1239
17. Sleegers K, Brouwers N, Van Damme P, Engelborghs S, Gijselinck I, van der Zee J et al (2009) Serum biomarker for progranulin-associated frontotemporal lobar degeneration. Ann Neurol 65:603–609
18. Finch N, Baker M, Crook R, Swanson K, Kuntz K, Surtees R et al (2009) Plasma progranulin levels predict progranulin mutation status in frontotemporal dementia patients and asymptomatic family members. Brain 132:583–591
19. Jian J, Konopka J, Liu C (2012) Insights into the role of progranulin in immunity, infection, and inflammation. J Leukoc Biol 93:199–208
20. Tang W, Lu Y, Tian QY, Zhang Y, Guo FJ, Liu GY et al (2011) The growth factor progranulin binds to TNF receptors and is therapeutic against inflammatory arthritis in mice. Science 332:478–484
21. Puga I, Cols M, Barra CM, He B, Cassis L, Gentile M et al (2012) B cell-helper neutrophils stimulate the diversification and production of immunoglobulin in the marginal zone of the spleen. Nat Immunol 13:170–180
22. Park B, Buti L, Lee S, Matsuwaki T, Spooner E, Brinkmann MM et al (2011) Granulin is a soluble cofactor for toll-like receptor 9 signaling. Immunity 34:505–513
23. Tian QY, Zhao YP, Liu CJ (2012) Modified yeast-two-hybrid system to identify proteins interacting with the growth factor progranulin. J Vis Exp 59:e3562
24. Feng JQ, Guo FJ, Jiang BC, Zhang Y, Frenkel S, Wang DW et al (2010) Granulin epithelin precursor: a bone morphogenic protein 2-inducible growth factor that activates Erk1/2 signaling and JunB transcription factor in chondrogenesis. FASEB J 24:1879–1892
25. Songsrirote K, Li Z, Ashford D, Bateman A, Thomas-Oates J (2010) Development and application of mass spectrometric methods for the analysis of progranulin N-glycosylation. J Proteomics 73:1479–1490
26. Gonzalez EM, Mongiat M, Slater SJ, Baffa R, Iozzo RV (2003) A novel interaction between perlecan protein core and progranulin: potential effects on tumor growth. J Biol Chem 278:38113–38116
27. Whitelock JM, Graham LD, Melrose J, Murdoch AD, Iozzo RV, Underwood PA (1999) Human perlecan immunopurified from different endothelial cell sources has different adhesive properties for vascular cells. Matrix Biol 18:163–178
28. Knox S, Merry C, Stringer S, Melrose J, Whitelock J (2002) Not all perlecans are created equal: interactions with fibroblast growth factor (FGF) 2 and FGF receptors. J Biol Chem 277:14657–14665
29. Knox S, Fosang AJ, Last K, Melrose J, Whitelock J (2005) Perlecan from human epithelial cells is a hybrid heparan/chondroitin/keratan sulfate proteoglycan. FEBS Lett 579: 5019–5023

Chapter 15

Compartment-Specific Flow Cytometry for the Analysis of TNF-Mediated Recruitment and Activation of Glomerular Leukocytes in Murine Kidneys

Volker Vielhauer

Abstract

Cytokines of the TNF superfamily, particularly TNF itself, are important mediators of inflammatory leukocyte recruitment and activation in parenchymal organs. In inflammatory kidney diseases, leukocytes accumulate in glomeruli and the tubulointerstitium, leading to glomerulonephritis and tubulointerstitial nephritis, respectively. In particular, glomeruli can be the target of organ-threatening leukocyte-mediated inflammation. As microvasculatures of the glomerulus and the tubulointerstitium differ markedly in their structural and functional properties, recruitment and subsequent activation of leukocytes to these sites occur via distinct mechanisms. To understand the pathways and mediators of leukocyte-driven inflammation in the kidney it is therefore essential to analyze glomerular and tubulointerstitial leukocyte recruitment in a compartment-specific way. The protocol presented here describes an easy and rapid technique that allows compartment-specific quantitation and qualitative analysis of leukocytes present in glomeruli and tubulointerstitial tissue by flow cytometry after separation of these tissue compartments.

Key words Glomerulonephritis, Interstitial nephritis, Immune cell subsets, Mouse

1 Introduction

Accumulation of leukocytes in glomeruli and tubulointerstitial tissue characterizes inflammatory kidney disease like glomerulonephritis and tubulointerstitial nephritis. Cytokine-induced upregulation of adhesion molecules and chemokines is essential for leukocyte–endothelial cell interactions, which are required for the recruitment of immune cells to renal tissue. In general, leukocyte recruitment occurs predominantly through interactions with post-capillary venules, with venular endothelial cells expressing appropriate adhesion molecules and presenting required chemokines to leukocytes. The renal circulation, however, is unusual in that it contains two separate capillary beds: the glomerular and peritubular capillary network. In contrast to peritubular capillaries, the microvascular bed of the glomerulus differs markedly from conventional

Jagadeesh Bayry (ed.), *The TNF Superfamily: Methods and Protocols*, Methods in Molecular Biology, vol. 1155, DOI 10.1007/978-1-4939-0669-7_15, © Springer Science+Business Media New York 2014

microvasculature in its structural and functional properties. Most prominently, glomerular capillaries form the only arterialized capillary network of the body, with the intracapillary pressure being about 40–50 mmHg. In contrast, the pressure in the peritubular capillaries of 5–10 mmHg is similar to that in capillary beds elsewhere. Not surprisingly, mechanisms of recruitment and properties of accumulating leukocyte subsets differ substantially in glomeruli and tubulointerstitial renal tissue. Recent advances in intravital imaging techniques have been particularly useful in revealing new leukocyte recruitment paradigms in glomeruli [1].

Understanding the mechanism of inflammatory responses in the kidney therefore requires compartment-specific studies of glomeruli and tubulointerstitium to identify cellular and humoral mediators with pathogenic relevance. The following protocol utilizes a previously published magnetic bead-based isolation technique for analysis of glomerular and tubulointerstitial tissue of individual mouse kidneys [2]. The isolated tissue is used for compartment-specific flow cytometry studies to simultaneously quantitate and characterize leukocyte populations in both renal compartments. Compartment specificity of the described technique was validated by analysis of different renal disease models with prominent infiltration of leukocytes either in the glomerular (nephrotoxic serum nephritis as a model of immune complex-mediated glomerulonephritis) or tubulointerstitial compartment (interstitial nephritis following unilateral ureteral obstruction), the latter with absent glomerular pathology [3]. In addition, mRNA and protein expression can be easily analyzed in the obtained glomerular and tubulointerstitial tissue fractions. Until recently, such compartment-specific studies relied on morphometric evaluation of immunohistochemically stained tissue sections and analysis of microdissected tissue specimens. The novel technique described here allowed us to demonstrate the constitutive presence of dendritic cells in normal mouse glomeruli, their TNF receptor-dependent accumulation, and their activation upon TNF exposure [3].

2 Materials

2.1 Magnetic Bead-Based Separation of Glomerular and Tubulointerstitial Tissue

1. Scissors, forceps, corked surface, autoclaved glass plate.
2. Custom-made pressure-controlled perfusion device with attached 20-gauge needle (*see* Fig. 1). Tube and needle may be sterilized in 70 % ethanol over night before use.
3. Magnetic particle concentrator (Life Technologies, Grand Island, NY, USA).
4. Water bath, cell culture microscope, Neubauer chamber.
5. 22-gauge needles, 1 mL syringes, 10 mL syringe pestle.

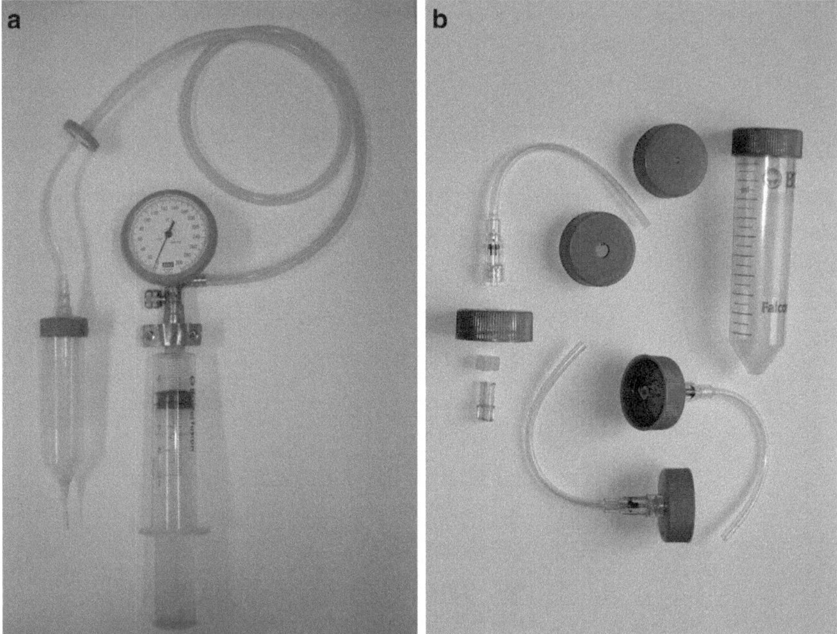

Fig. 1 Setup of the pressure-controlled perfusion device. (**a**) A 50 mL plastic syringe is connected to a commercial blood pressure manometer via a short flexible plastic tube. Syringe and manometer are stabilized by a small steel plate. Instead of a blood pressure cuff a 50 mL conical polypropylene tube is connected to the manometer through a flexible plastic tube, a 0.2 μm syringe filter, and a plastic infusion line. (**b**) Fixed with a plastic adhesive, the infusion line directly inserts into a hole in the screw cap of the tube. The bottom of the 50 mL tube is cut open to attach a 20-gauge Luer-lock needle using plastic adhesive or dental cement. Before perfusion the plunger of the 50 mL syringe is almost completely retracted. The 50 mL tube is positioned upright, the needle placed in a plastic needle cap, the tube filled with 40 mL of PBS and 200 μL of Dynabeads, the screw cap with attached infusion line placed on the tube, and a slight negative pressure is applied by pulling back the plunger of the 50 mL syringe a few millimeters to prevent outflow of the Dynabead suspension. During perfusion constant pressure is applied to the system by pushing the plunger of the 50 mL syringe

6. 1.5 mL microcentrifuge tubes, 15 mL tubes, 50 mL tubes.
7. Pasteur pipettes.
8. 100 μm cell strainer (BD Biosciences, MA).
9. Anesthetic: medetomidine-midazolam-fentanyl mix (0.5/5/0.05 mg/kg body weight).
10. M-450 tosylactivated Dynabeads, 4.5 μm diameter (Life Technologies).
11. Buffer 1: 0.1 M sodium phosphate buffer, pH 8.0 (mix 93.2 mL of 1 M Na_2HPO_4 with 6.8 mL of 1 M NaH_2PO_4 and dilute with H_2O to final volume of 1 L).
12. Blocking solution: buffer 1 supplemented with 0.5 % nuclease-free bovine serum albumin (BSA).

13. Buffer 2: Ca^{2+} and Mg^{2+} free phosphate buffered saline (PBS), pH 7.4.
14. Collagenase A (Roche Diagnostics, Mannheim, Germany).
15. Sterile PBS, Hanks' balanced salt solution (HBSS).

2.2 Preparation of Glomerular and Tubulointerstitial Single Cell Suspensions

1. Waterbath, vortex mixer, Neubauer chamber.
2. 2 mL microcentrifuge tubes with round bottom, 50 mL tubes, 15 mL tubes, FACS tubes.
3. Pasteur pipettes.
4. 60 mm untreated plastic petri dishes.
5. 19-gauge needles, 27-gauge needles, 30-gauge needles, 10 mL syringes.
6. 70 μm cell strainer (BD Biosciences).
7. Collagenase type I (Sigma-Aldrich, MO), deoxyribonuclease type I (Sigma-Aldrich).
8. Sterile PBS, HBSS with Ca^{2+} and Mg^{2+}.
9. HBSS without Ca^{2+} and Mg^{2+}-0.2 M ethylenediaminetetraacetic acid (EDTA).
10. FACS buffer: 0.2 % (w/v) BSA, 0.1 % (w/v) sodium azide in PBS.

2.3 Quantitation and Characterization of Glomerular and Tubulointerstitial Leukocyte Populations by Multicolor Flow Cytometry

1. Flow cytometer (e.g., FACS Calibur™ flow cytometer, BD Biosciences), flow cytometry analysis software (e.g., Cellquest™).
2. Vortex mixer.
3. FACS tubes.
4. FACS buffer: 0.2 % (w/v) BSA, 0.1 % (w/v) sodium azide in PBS.
5. FACS blocking buffer: FACS buffer, 10 % normal rat serum.
6. Fluorochrome-conjugated antibodies to leukocyte surface antigens (e.g., from BD Biosciences).

3 Methods

3.1 Magnetic Bead-Based Separation of Glomerular and Tubulointerstitial Tissue: Preparation of Dynabeads

1. Prepare 200 μL of M-450 tosylactivated Dynabeads (8×10^7 beads) for each mouse to be perfused.
2. Resuspend the beads in the original vial by vortexing for >30 s.
3. Transfer the appropriate volume of beads to a 1.5 mL microcentrifuge tube.
4. Place the tube in the magnetic particle concentrator for 1 min, aspirate and discard the supernatant, and remove the tube from the magnet.

5. Resuspend the washed beads in the same volume of buffer 1 as the initial volume of beads (200 μL per mouse), or at least 1 mL, by pipetting solution up and down for 2 min.
6. Repeat **steps 4** and **5** two times, finally resuspending beads in the same volume of buffer 1 as the initial volume of beads (200 μL per mouse).
7. Add the same volume (200 μL per mouse) blocking solution and mix by pipetting solution up and down.
8. Incubate beads for 16–24 h at 4 °C.
9. Place the tube in the magnet for 1 min, aspirate and discard the supernatant, and remove the tube.
10. Resuspend the blocked beads with same volume (200 μL per mouse) of buffer 2 by pipetting solution up and down.
11. Repeat **steps 9** and **10** two times, finally resuspending beads in the same volume of buffer 2 as the initial volume of beads (200 μL per mouse).
12. Blocked beads resuspended in buffer 2 can be stored at 4 °C for several months.

3.2 Magnetic Bead-Based Separation of Glomerular and Tubulointerstitial Tissue: Perfusion of Mouse Kidneys with Dynabeads

1. Anesthetize the mouse according to local standards, e.g., by intraperitoneal injection of a medetomidine-midazolam-fentanyl mix (0.5/5/0.05 mg/kg body weight) with a 1 mL syringe and 22-gauge needle.
2. To ensure that the mouse is properly sedated, pinch the toes to judge its level of response to a painful stimulus, which should be absent.
3. Place the sedated mouse on a corked surface with the abdomen facing up. With small needles secure the four paws to the surface spreading them as wide as possible. Wet the abdominal fur with 70 % ethanol.
4. Grab the skin with the forceps at the level of the diaphragm and cut open the abdominal wall by a longitudinal incision to expose the liver. Cut laterally and then up, cutting through the ribs. Lift flap and continue cutting until the heart is assessable. Secure chest flap to corked surface with a needle.
5. Place the 50 mL tube of the perfusion device with attached 20-gauge needle (*see* Fig. 1) in an upright position and block the needle by placing it in a plastic needle cap. Fill tube with 40 mL of sterile PBS, prewarmed to 37 °C in a waterbath. Add 200 μL of blocked Dynabeads and mix carefully by pipetting up and down. Check that no air bubbles form within the tube and needle. Fix the screw cap with attached wiring to the tube and apply a slight negative pressure to prevent outflow of the Dynabead suspension.

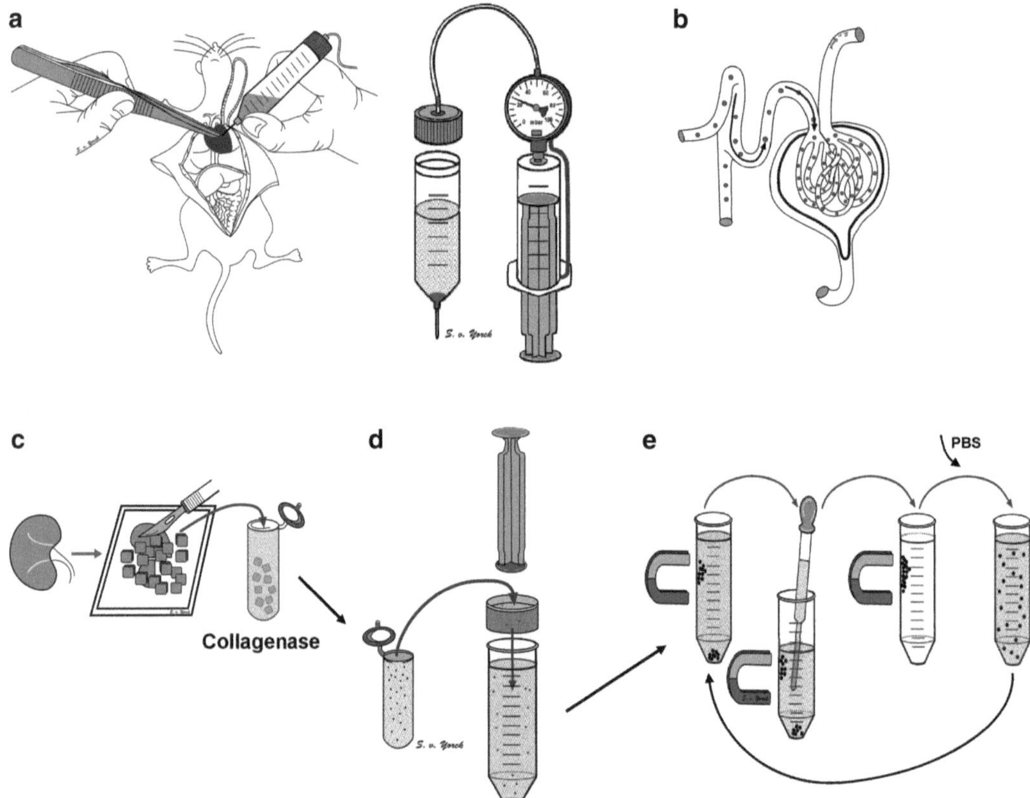

Fig. 2 Flow chart of the magnetic isolation technique of mouse glomeruli and tubulointerstitial tissue. (**a**) Perfusion of Dynabeads through the left ventricle with the pressure-controlled perfusion device. (**b**) Trapping of Dynabeads in glomerular capillaries. (**c**) Removing kidneys, mincing into 1 mm^3 pieces, and digestion with collagenase. (**d**) Filtration of digested tissue through 100 μm cell strainer. (**e**) Separation of glomeruli by a magnetic particle concentrator and washing

6. Insert 20-gauge needle of the perfusion device into the left cardiac ventricle (*see* Fig. 2a).
7. A second person should start perfusion of the Dynabeads by initially applying a pressure of 20 mmHg for 20 s. Immediately after starting the perfusion cut the inferior vena cava below the right atrium.
8. Perfuse the complete volume of PBS with suspended Dynabeads (40 mL) applying a constant pressure of 40 mmHg. A good perfusion is indicated by a color change of liver and kidneys into a light coffee color.
9. Remove perfusion device and harvest both kidneys by cutting the renal pedicle. Remove renal capsule by gently squeezing kidneys between two fingers. Place kidneys on an autoclaved glass plate on ice.

3.3 Magnetic Bead-Based Separation of Glomerular and Tubulointerstitial Tissue: Isolation of Tissue Fractions

Due to the unique vascular architecture of the kidney, the perfused Dynabeads are selectively targeted into glomeruli (*see* Fig. 2b), with a minor amount reaching the peritubular capillary network [2]. Incubation of perfused renal tissue with collagenase A digests tubulointerstitial tissue liberating magnetic beads in interstitial capillaries. In contrast, glomeruli containing trapped beads remain intact due to the relative resistance of the Bowman's capsule to collagenase digestion. This allows separation of magnetically labeled intact glomeruli from digested tubulointerstitial cells. With the exception of the collagenase digestion all of the following steps are performed on ice.

1. With the kidneys placed on the glass plate on ice, mince tissue with a scalpel into 1 mm^3 pieces. Transfer tissue from each kidney into a separate 2 mL microcentrifuge tube with round bottom (*see* Fig. 2c and **Note 1**).
2. Add 1 mL of prewarmed 1 mg/mL collagenase A dissolved in HBSS (with Ca^{2+} and Mg^{2+}). Incubate for 30 min at 37 °C in a water bath. Do not vortex or shake, but occasionally mix solution by gentle agitation during the incubation (*see* Fig. 2c).
3. Gently press digested tissue through a 100 μm cell strainer placed on a 50 mL tube on ice using a flattened syringe plunger or pestle (*see* Fig. 2d). Wash the 2 mL microcentrifuge tube with 1 mL of PBS, 4 °C, and pass solution through the cell strainer. Wash cell strainer two times with 1 mL of PBS (including the lower surface).
4. Replace 100 μm cell strainer on the 50 mL tube containing digested tissue of the first kidney. Repeat **step 3** with the second kidney of the same mouse, combining all digested renal tissue of one animal (*see* **Note 2**).
5. Transfer digested tissue (total volume is approximately 8 mL) into a cooled 15 mL tube with round bottom. Mix by gentle agitation and store on ice.
6. Place 15 mL tube with magnetically labeled glomeruli in the magnetic particle concentrator for 7 min. Keep tube on ice during this step (*see* Fig. 2e).
7. Magnetically labeled glomeruli and free Dynabeads will attach at the lateral side of the 15 mL tube placed in the magnet. Using a Pasteur pipette carefully aspirate the supernatant without touching the attached glomeruli with the tip of the pipette (*see* Fig. 2e).
8. Store the first supernatant on ice until further use. This supernatant contains tubular fragments and tubulointerstitial cells, but no glomeruli (*see* Fig. 3a).
9. Remove the tube from the magnet and resuspend the remaining tissue in 10 mL of PBS, 4 °C, by gentle agitation (*see* Fig. 2e). Do not vortex or shake.

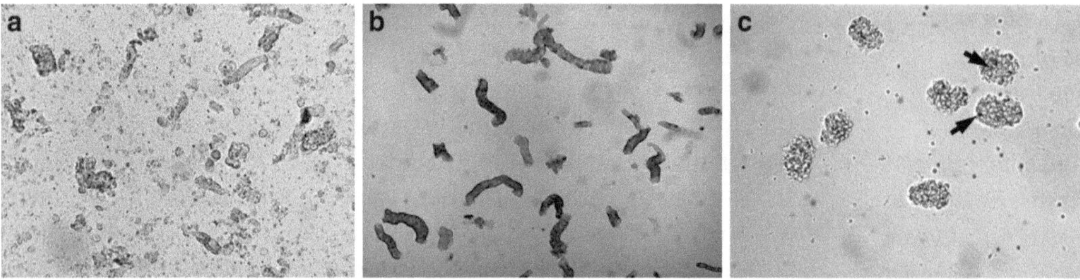

Fig. 3 Renal tissue fractions after magnetic bead-based separation. (**a**) Tubulointerstitial tissue fraction (supernatant 1) containing tubular fragments, tubular cells, leukocytes, and polymorphic interstitial cells, but no glomeruli. (**b**) Tubular fraction (supernatant 2) lacking glomeruli, interstitial cells, and leukocytes. (**c**) Isolated glomeruli. *Arrowheads* indicate magnetic Dynabeads within glomerular capillaries. Original magnification ×100 (reproduced from [3] with permission from Nature Publishing Group)

10. Place the tube again in the magnet for 5 min. Keep tube on ice during this step.
11. Aspirate the supernatant as in **step** 7.
12. Store the second supernatant on ice until further use. The second supernatant contains tubular fragments only, but no other tubulointerstitial cells or glomeruli (*see* Fig. 3b).
13. Remove the tube from the magnet and resuspend tissue in 4 mL of PBS, 4 °C, by gentle agitation.
14. Filter the tissue suspension through a 100 μm cell strainer into a 50 mL tube. This time do not use a plunger to pass tissue through the filter. Combine filtrate with two subsequent washes of the cell strainer with 1 mL of PBS, 4 °C.
15. Transfer tissue (total volume is approximately 6 mL) into a cooled 15 mL tube with round bottom. Mix by gentle agitation and store on ice.
16. Place 15 mL tube with glomeruli in the magnetic particle concentrator for 5 min. Keep tube on ice during this step.
17. Aspirate the supernatant as in **step** 7.
18. Remove the tube from the magnet and resuspend the remaining tissue in 10 mL of PBS, 4 °C, by gentle agitation. Do not vortex or shake.
19. Wash glomeruli two more times by repeating **steps 16–18**.
20. Finally, resuspend washed glomeruli with 1.5 mL of PBS, 4 °C, and transfer suspension into a 1.5 mL microcentrifuge tube.
21. Place 1.5 mL tube in the magnetic particle concentrator for 5 min. Keep tube on ice.
22. Aspirate the supernatant as in **step** 7 and check under a cell culture microscope for any remaining cells and cell debris.
23. Repeat **steps 20–22** until supernatant is completely free of any contaminating cell debris.

24. Finally, check purity and integrity of the glomerular preparation under the microscope (*see* Fig. 3c and **Notes 3–5**). Determine the number of isolated glomeruli by counting a 10 μL aliquot of resuspended glomeruli in a Neubauer chamber (*see* **Notes 6–9**).

3.4 Preparation of Glomerular and Tubulointerstitial Single Cell Suspensions

Single cell suspensions (*see* **Note 10**) from glomerular, tubulointerstitial (first supernatant), and tubular preparations (second supernatant) are prepared essentially as from whole kidney [4].

1. Transfer separated tissue fractions (glomeruli, tubulointerstitial, and tubular tissue) into 15 mL tubes, add 10 mL of HBSS (with Ca^{2+} and Mg^{2+}), 4 °C, centrifuge at 250×*g*, 4 °C for 5 min, and aspirate the supernatant.
2. Resuspend tissue pellets in 5 mL of prewarmed HBSS (with Ca^{2+} and Mg^{2+}), containing 1 mg/mL collagenase type I and 0.1 mg/mL deoxyribonuclease type I. Incubate at 37 °C in the water bath for 20 min. Mix samples every 2 min by vigorous shaking.
3. Centrifuge samples at 250×*g*, 4 °C for 5 min, aspirate supernatant.
4. Resuspend pellet in 10 mL of HBSS (with Ca^{2+} and Mg^{2+}), 4 °C for 5 min by vortexing.
5. Wash pellet a second time repeating **steps 3** and **4**.
6. Resuspend pellet in 5 mL of prewarmed HBSS (without Ca^{2+} and Mg^{2+}) containing 2 mM of EDTA and incubate at 37 °C in the water bath for 20 min. Mix samples occasionally by shaking.
7. Centrifuge samples at low speed (!) with 30×*g* at 4 °C for 5 min, carefully aspirate supernatant using a Pasteur pipette, and transfer supernatant with single cells into a 15 mL tube placed on ice until further use (*see* **Note 11**).
8. Digest the remaining tissue pellet a second time adding 5 mL of prewarmed HBSS (with Ca^{2+} and Mg^{2+}) containing 1 mg/mL of collagenase type I. Incubate for 20 min, 37 °C in the water bath. Mix samples every 2 min by vigorous shaking.
9. Transfer the suspension into plastic petri dishes placed on ice.
10. Aspirate suspension using a 10 mL syringe and subsequently pass through a 19-gauge, 27-gauge, and 30-gauge needle. Finally, combine suspension with supernatant from **step** 7.
11. Centrifuge samples at 250×*g*, 4 °C for 5 min, aspirate the supernatant.
12. Resuspend pellet in 10 mL of ice-cold PBS, 4 °C.
13. Wash cell suspension a second time repeating **step 12**.
14. Resuspend pellet in 1 mL of FACS buffer, 4 °C.

15. Pre-wet a 70 μm cell strainer with 1 mL of FACS buffer (*see* **Note 12**). Gently press cell suspension through the cell strainer placed on a 50 mL tube on ice using a flattened syringe plunger. Combine filtrate with a subsequent wash of the cell strainer with 10 mL of PBS, 4 °C, and transfer suspension into a 15 mL tube.
16. Centrifuge samples at 250×*g*, 4 °C for 5 min, aspirate supernatant.
17. Resuspend pellet in 1 mL of FACS buffer, 4 °C, and store cell suspension on ice until further use.
18. Using a 10 μL aliquot, count cells in a Neubauer chamber.

3.5 Quantitation and Characterization of Glomerular and Tubulointerstitial Leukocyte Populations by Multicolor Flow Cytometry

1. Aliquot approximately 1×10^6 cells suspended in 100 μL of FACS buffer into FACS tubes and store on ice.
2. To block unspecific binding of fluorochrome-labeled antibodies to cells preincubate samples with 10 % normal rat serum at 4 °C for 5 min by adding 11 μL of serum to each sample (*see* **Note 13**).
3. Depending on the leukocyte populations and activation markers of interest stain cell suspensions with a combination of fluorochrome-labeled antibodies to leukocyte surface markers or respective isotype controls by adding appropriate amounts of antibody as recommended by the manufacturer (between 0.1 and 1.0 μg antibody per 100 μL for 10^6 cells). Mix by vortexing and incubate at 4 °C for 40 min in the dark.
4. Add 2 mL of FACS buffer, 4 °C to each sample and vortex.
5. Centrifuge samples at 250×*g*, 4 °C, aspirate supernatant and resuspend pellet in 2 mL of FACS buffer, 4 °C.
6. Wash again repeating **step 5**.
7. Finally, resuspend pellet in 300–500 μL of FACS buffer (final concentration of cells approximately 10^6 cells in 0.5 mL). Perform flow cytometry as soon as possible after staining (*see* **Note 14**).
8. An example for the gating strategy to identify glomerular leukocyte populations ($CD45^+$ leukocytes, $F4/80^+$ mononuclear phagocytes, $CD11c^+$ dendritic cells, $Ly6G^+$ neutrophils, $CD3^+$ $CD4^+$ T cells, and $CD3^+$ $CD8^+$ T cells) is illustrated in Fig. 4. Identified leukocyte subpopulations can be quantitated (*see* **Notes 15** and **16**) and further characterized by staining for appropriate activation markers [3].

4 Notes

1. A small portion of each kidney may be fixed in formalin and used for histology and immunohistochemical studies. With this approach the efficacy of the glomerular perfusion with

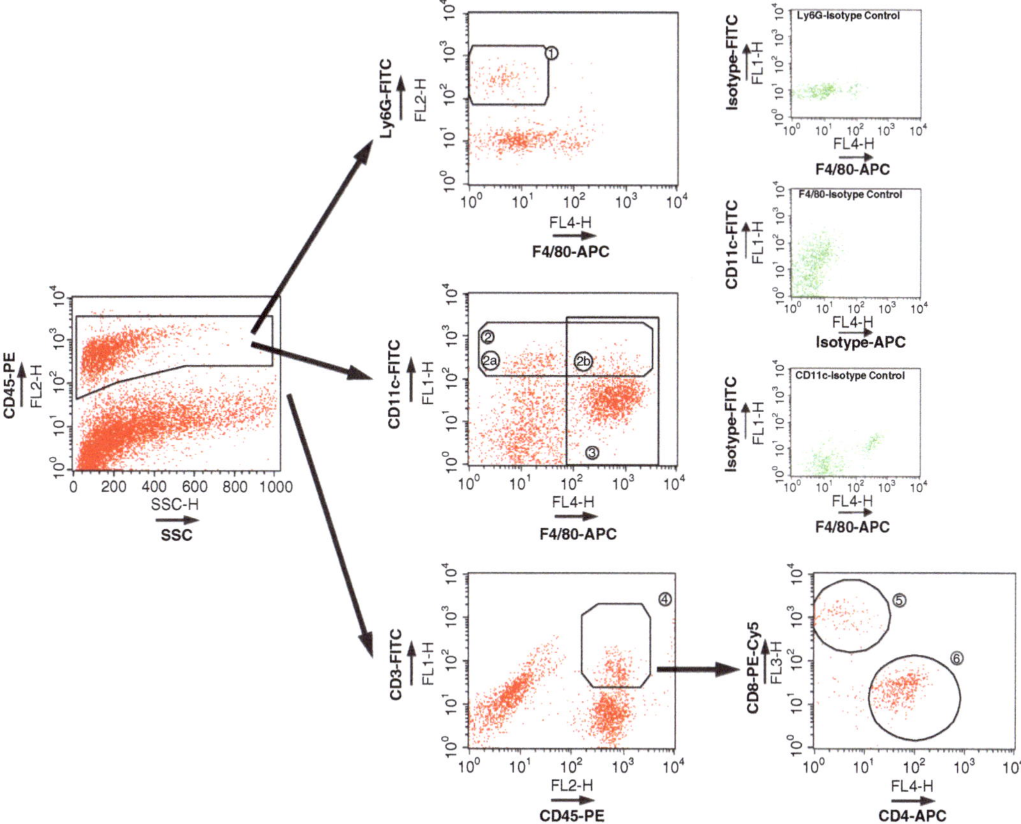

Fig. 4 Gating strategy for glomerular and interstitial leukocyte populations. The example of multicolor flow cytometry dot blots and gating shown is from isolated glomeruli of a nephritic mouse (day 7 after induction of nephrotoxic serum nephritis). Staining definitions for the individual cell populations were as follows: neutrophils (*1*): CD45$^+$ Ly6G$^+$ F4/80$^-$; F4/80 positive mononuclear phagocytes (*2a*): CD45$^+$ CD11c$^+$ F4/80$^+$; F4/80 negative mononuclear phagocytes (*2b*): CD45$^+$ CD11c$^+$ F4/80$^-$; macrophages-like mononuclear phagocytes (*3*): CD45$^+$ CD11c$^-$ F4/80$^+$; T cells (*4*): CD45$^+$ CD3ε^+; CD4 positive T cells: CD45$^+$ CD3ε^+ CD4$^+$ CD8$^-$ (*5*); and CD8 positive T cells (*6*): CD45$^+$ CD3ε^+ CD4$^-$ CD8$^+$ (reproduced from [3] with permission from Nature Publishing Group)

Dynabeads can also be monitored, as some beads should be visible in capillaries of most glomerular cross sections.

2. The described protocol can also be used to isolate and purify glomerular, tubulointerstitial, and tubular compartments of one single kidney. This is especially important when only one kidney per animal is subjected to a disease model, e.g., in unilateral ureteral obstruction [3] or unilateral ischemia-reperfusion injury.
3. After separation of glomerular and tubulointerstitial tissue the final magnetically labeled fraction contains highly purified glomeruli with a purity >97 % (intact glomeruli versus nonglomerular particles in the isolate).
4. As originally reported, isolated glomeruli lack the Bowman's capsule [2]. Occasionally parts of the afferent and efferent

arterioles are still attached. The absence of the Bowman's capsule in isolated glomeruli prevents contamination of the glomerular tissue preparation by periglomerular leukocyte infiltrates of the interstitium that may adhere to the capsule. This is of critical importance to the subsequent compartment-specific flow cytometric analysis of glomerular versus tubulointerstitial leukocyte populations.

5. Structurally damaged glomeruli may result from inappropriately long collagenase treatment, rigorous shaking during the isolation procedure, or crushing glomeruli during the sieving steps.
6. The number of glomeruli isolated with this protocol should reach a yield of approximately 20,000 glomeruli per mouse.
7. We obtained highest yields of glomeruli when 6–8 week old mice were perfused. However, younger and older mice may be used. In addition, the described technique to isolate and purify glomeruli from adult mice [2] has been successfully adapted to allow isolation of immature and mature glomeruli of embryonic mice [5].
8. Isolation of glomeruli from mice subjected to glomerular disease models is not feasible when advanced glomerular sclerosis or thrombosis is present that prevents efficient glomerular perfusion with Dynabeads (e.g., in later stages of nephrotoxic serum nephritis or adriamycin nephropathy). In this case isolation of glomeruli must be performed at early time points after induction of the disease.
9. If the yield of isolated glomeruli is low the supernatants obtained during the isolation procedure should be examined under a cell culture microscope:
   - Glomeruli with no detectable or low numbers of retained Dynabeads that are present in the first and second aspirated supernatant indicate insufficient perfusion and trapping of Dynabeads in glomeruli, as glomeruli without beads will not be retained by the magnet. The perfusion technique should be checked (applied pressure, prewarmed solution, correct puncture of the left ventricle?). Correct perfusion will result in 6–25 beads retained in each glomerulus.
   - Bead containing glomeruli present in aspirated supernatants indicate a loss of glomeruli during the isolation procedure. This is a hint for an inadequate pipette technique during aspiration with disturbance of the glomerular pellet at the lateral wall of the tube.
10. Isolated glomeruli can be directly used for subsequent compartment-specific flow cytometry. Alternatively, glomeruli can be transferred into RLT buffer or RNA-Later for subsequent mRNA isolation and expression studies, or into

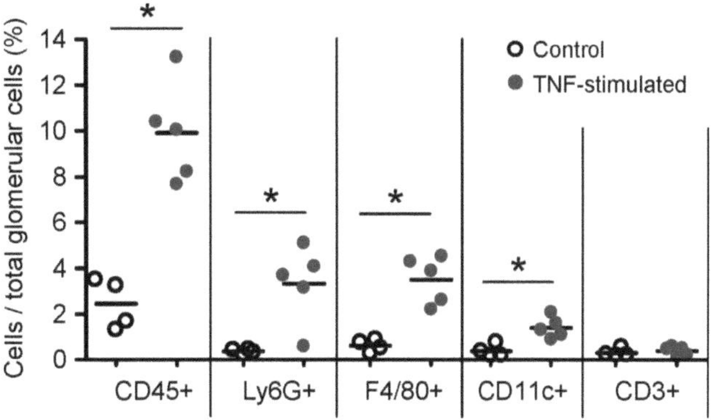

Fig. 5 TNF-dependent accumulation of glomerular leukocytes. Compartment-specific flow cytometry was performed on glomerular tissue prepared from wildtype mice 8 h after intraperitoneal injection of 5 μg TNF. Numbers of CD45$^+$ leukocytes, Ly6G$^+$ granulocytes, F4/80$^+$ mononuclear phagocytes, CD11c$^+$ dendritic cells, and CD3$^+$ T cells are expressed as percentage of all glomerular cells. *$p < 0.05$ versus wildtype (adapted from [6] with permission from PLOS)

appropriate lysing buffers for analysis of protein expression by Western blot or ELISA.

11. Centrifugation with low *g* is of critical importance in this step. Otherwise single cells will not remain in the supernatant.
12. Without prewetting the cell strainer with FACS buffer recovery of single cells will be reduced due to increased clumping of cells on the filter surface.
13. To reduce background of fluorochrome-labeled antibody staining for flow cytometry FcγII/III receptor-mediated antibody binding can be additionally blocked by preincubation of cell suspensions with unlabeled anti-mouse CD32/CD16 antibodies (Fc block). Approximately 0.25 μg of Fc block can be added to 1×10^6 cells for 5 min, at 4 °C before incubation with 10 % normal rat serum and need not be washed out prior to addition of fluorochrome-labeled antibodies.
14. Stained samples may be stored in FACS buffer at 4 °C in the dark without fixation for up to 18 h post-staining before analysis, without loss of viability and surface staining.
15. We routinely perform compartment-specific quantitation of leukocyte subsets by relating the number of leukocytes to the total number of cells within the analyzed sample (*see* Fig. 5). The latter is represented by the total event rate of the FACS analysis using an appropriate constant threshold of the side scatter.
16. Alternatively, absolute leukocyte counts may be determined by adding counting beads to isolated glomerular and tubulointerstitial samples of both kidneys before preparing single cell

suspensions of the samples (*see* **step 1** of Subheading 3.4). As the absolute number of isolated glomeruli from one mouse (or one kidney) is known (*see* **step 24** of Subheading 3.3), the number of leukocytes per glomerulus can also be calculated. However, due to the relative resistance of glomeruli to collagenase digestion glomerular tissue may not be completely digested into a single cell suspension by incubation with collagenase, deoxyribonuclease, EDTA, and subsequent mechanical shearing (*see* Subheading 3.4). Leukocytes contained in undigested portions of glomeruli, which are retained in the cell strainer after filtration, will not be counted. This will result in an underestimated number of glomerular leukocytes actually present in vivo. If absolute quantitation of glomerular leukocytes is performed, the glomerular single cell preparations should therefore be carefully checked for the proportion of undigested glomerular fragments under the microscope.

Acknowledgements

The author thanks Andreas Blutke, Institute for Veterinary Pathology, Ludwig-Maximilians-University Munich for sharing his experience with the magnetic glomerular isolation procedure, Werner Vielhauer for designing the pressure-controlled perfusion device, and Nuru Eltrich for his continued technical support. Sibylle von Yorck is acknowledged for providing the excellent graphical illustration of the glomerular isolation procedure. This work was supported by DFG grant VI 231/2-1 to V. V.

References

1. Devi S, Li A, Westhorpe CL, Lo CY, Abeynaike LD, Snelgrove SL et al (2013) Multiphoton imaging reveals a new leukocyte recruitment paradigm in the glomerulus. Nat Med 19: 107–112
2. Takemoto M, Asker N, Gerhardt H, Lundkvist A, Johansson BR, Saito Y et al (2002) A new method for large scale isolation of kidney glomeruli from mice. Am J Pathol 161:799–805
3. Schwarz M, Taubitz A, Eltrich N, Mulay SR, Allam R, Vielhauer V (2013) Analysis of TNF-mediated recruitment and activation of glomerular dendritic cells in mouse kidneys by compartment-specific flow cytometry. Kidney Int 84:116–129
4. Vielhauer V, Allam R, Lindenmeyer MT, Cohen CD, Draganovici D, Mandelbaum J et al (2009) Efficient renal recruitment of macrophages and T cells in mice lacking the duffy antigen/receptor for chemokines. Am J Pathol 175:119–131
5. Cui S, Li C, Ema M, Weinstein J, Quaggin SE (2005) Rapid isolation of glomeruli coupled with gene expression profiling identifies downstream targets in Pod1 knockout mice. J Am Soc Nephrol 16:3247–3255
6. Taubitz A, Schwarz M, Eltrich N, Lindenmeyer MT, Vielhauer V (2013) Distinct contributions of TNF receptor 1 and 2 to TNF-induced glomerular inflammation in mice. PLoS One 8:e68167

Chapter 16

Determination of Soluble Tumor Necrosis Factor Receptor 2 Produced by Alternative Splicing

Xavier Romero, Juan D. Cañete, and Pablo Engel

Abstract

Soluble cytokine receptors have proven to be very useful biomarkers in a large variety of diseases, including cancer, infections, and chronic inflammatory diseases. These soluble receptors are produced by proteolytic cleavage or alternative splicing. Several cytokine receptors including tumor necrosis factor receptor 2 (TNFR2) can be generated by both mechanisms. However, the conventional ELISA systems do not differentiate between these two types of soluble receptors. We describe a sandwich ELISA to specifically quantify soluble TNFR2 protein generated by alternative splicing. This method requires the use of a capturing monoclonal antibody (mAb) specific of an epitope present in the soluble TNFR2 generated by alternatively splicing but absent in the proteolytically generated isoform. Here we present a detailed protocol for the production and validation of such a mAb. This method has the potential to be applied for measuring other soluble cell surface molecules generated by alternative splicing.

Key words Soluble cytokine receptor, Splice variant, Monoclonal antibodies, TNFR2

1 Introduction

Soluble forms of many membrane-linked proteins, including cytokine receptors, have been described [1, 2]. Soluble cytokine receptors bind their ligands with comparable affinity and are regulators of inflammation, cellular proliferation, and apoptosis. Most of these soluble proteins function as decoy receptors inhibiting cytokine activity by competing with their membrane-associated counterparts and preventing cell signaling. However, certain soluble receptors have been shown to be agonists, promoting cytokine activity [2].

Soluble receptors, which are produced by both normal and cancer cells, can be generated by two distinct molecular mechanisms. First, proteolytic protein cleavage, also known as shedding, can lead to a soluble receptor ectodomains. The second mechanism consists of alternative splicing of mRNA transcripts that selectively deletes the transmembrane domain of membrane-associated receptors (Fig. 1). Interestingly, several cytokine receptors such as

Jagadeesh Bayry (ed.), *The TNF Superfamily: Methods and Protocols*, Methods in Molecular Biology, vol. 1155, DOI 10.1007/978-1-4939-0669-7_16, © Springer Science+Business Media New York 2014

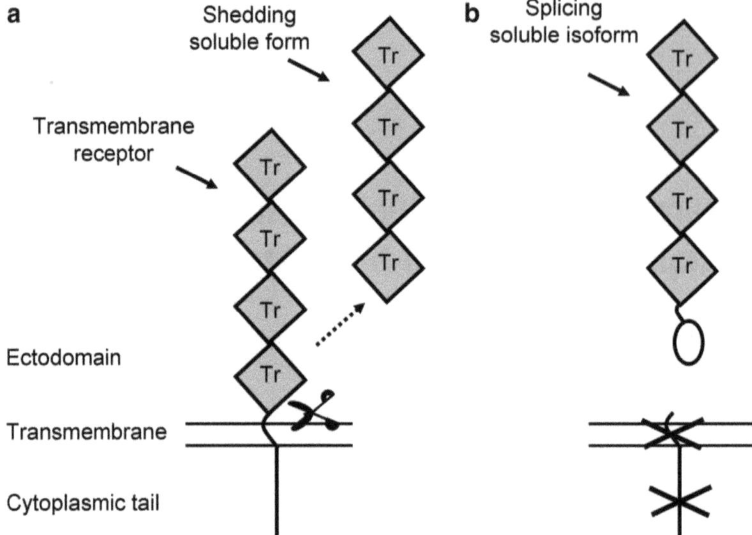

Fig. 1 Mechanisms of generation of soluble cell surface receptors. (**a**) The transmembrane isoform of the receptors consists of extracellular domain (Ex), transmembrane (TM), and cytoplasm (Cyto) domains. Soluble receptors can be produced by a proteolytic cleavage, also named shedding or (**b**) by alternative splicing of mRNA transcript. The latter process is produced by skipping one or more exons which in some cases leads to a translation reading frame shift and termination due to a premature stop codon. Tr—TNF receptor domain

Table 1
Soluble proteins generated by proteolytic cleavage (shedding), alternative splicing, or both mechanisms

Proteolytic cleavage (PC)	Alternative splicing (AS)	Both PC and AS
IL-2Rα	IL-4Rα	IL1-RII
TGF-βRIII	IL-5Rα	IL-6Rα
M-CSFR	IL-7Rα	IL-15Rα
c-kit (CD117)	IL-9Rα	TNFRSF1B (TNFR2)
PDGFR	LIFRα	
TNFRSF8 (CD30)	GM-CSFRα	
TNFRSF1A (TNFR1)	TNFRSF6 (CD95, Fas)	

IL-1RII, IL-6Rα, IL-15Rα, and tumor necrosis factor receptor 2 (TNFR2, TNFRSF1B) can be generated by both mechanisms: proteolytic cleavage and alternative splicing [3] (Table 1).

Elevated levels of the soluble cytokine receptors have been identified in human serum in a wide variety of inflammatory and neoplastic diseases. Some are useful as disease activity biomarkers, whereas others may directly contribute to the pathogenesis of the disease process [4].

Soluble TNFRs play a critical role in regulating the inflammatory response by binding and eventually neutralizing free TNF-α [5]. Elevated levels of soluble TNFR2 that correlate with clinical features and outcome have been observed in the serum of patients with cancer, chronic inflammatory pathologies such as type 1 and type 2 diabetes, and autoimmune diseases [6–10]. Very high levels have been detected in sepsis where they correlated with the clinical stages and disease progression [11]. We have demonstrated that the predominant form of soluble TNFR2 found in patients with sepsis is generated by shedding, whereas patients with rheumatoid arthritis present high levels of soluble TNFR2 produced by alternative splicing that lacks the transmembrane and part of the cytoplasmic domains [12]. Moreover, sTNFR2 levels generated by alternative splicing may influence the response to anti-TNF therapy and the joint damage in patients with rheumatoid arthritis [13]. No correlation between the levels of soluble TNFR2 produced by the two mechanisms is observed, indicating independent regulatory pathways of soluble TNFR2 release [12].

Soluble cytokine receptors are commonly determined by sandwich ELISA using antibodies against two different epitopes located in the extracellular domain. ELISA kits that quantify almost all soluble receptors can be purchased from several vendors. However, these ELISA kits are not able to distinguish between the soluble receptors generated by shedding or alternative splicing.

We present a procedure for specifically determining soluble cytokine receptor generated by alternative splicing. Moreover this protocol allows differentiation of these soluble forms from those generated by shedding. This technique is based on a sandwich ELISA in which the capturing monoclonal antibody recognizes a sequence only present in the splice variant (Fig. 2).

The epitope recognized by the capturing antibody has to correspond to a sequence present in the cytoplasmic tail or ideally to a unique sequence generated by the alternative spliced mechanism. In most cases alternative splicing can result in a reading frame shift that results in soluble proteins with distinct carboxyl-termini.

Here we present a protocol to quantify soluble TNFR2 generated by alternative splicing, but this same procedure can be applied to develop methods to determine other alternatively spliced soluble cell surface receptors using mAbs directed against sequences unique to these proteins. Note that every research group uses its own methodology to generate mAbs. Here, we present a protocol that generates fast results, does not require medium changes, avoiding the risk of contaminations, and guarantees the specificity of the antibodies.

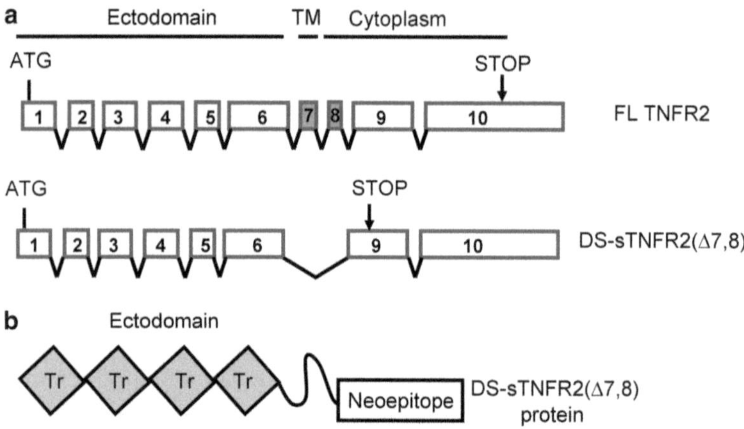

Fig. 2 TNFR2 soluble isoform generated by alternative splicing. (**a**) Exon organization of the full length (FL TNFR2) and a soluble splicing isoform (DS-sTNFR2 (Δ7,8)) of TNFR2. Protein ectodomain is encoded by exons 1–6, transmembrane by exon 7, and the cytoplasmic tail by exons 8–10. The soluble splicing form is generated by skipping exons 7 and 8, which results in a reading frame shift and a premature STOP codon. (**b**) As a consequence of the reading frame shift, a soluble isoform (DS-sTNFR2 (Δ7,8)) containing a unique C-terminal amino acid sequence (Ala-Ser-Leu-Ala-Cys-Arg) is generated. Tr—TNF receptor domain, ATG—Start codon, STOP—Stop codon

2 Materials

1. 16–20 week old BALB/c mice.
2. NS1 myeloma cells (mouse B line) (European Collection of Cell Cultures, Salisbury, UK) (*see* **Note 1**).
3. COS-7 cells (American Type Culture Collection, Manassas, VA).
4. Synthetic Peptide: PMGPSPPAEGSTGD.
5. Complete Freund's adjuvant (CFA, SIGMA, St Louis, MO), incomplete Freund's adjuvant (IFA, SIGMA).
6. RPMI complete medium: RPMI medium supplemented with 10 % preselected fetal calf serum (FCS) (*see* **Note 1**), 4 mM L-glutamine, 50 U penicillin, 50 μg streptomycin.
7. HAT medium: 500 mL RPMI, 2.5 mL Hybridoma fusion and cloning supplement (ROCHE), 50× concentrate of HAT (SIGMA), 20 % preselected FCS, 4 mM L-glutamine, 50 U penicillin, 50 μg streptomycin.
8. HT medium: 500 mL RPMI, 2.5 mL Hybridoma fusion and cloning supplement, 50× concentrate of HT, 20 % preselected FCS, 4 mM L-glutamine, 50 U penicillin, 50 μg streptomycin.
9. Polyethylene glycol solution—PEG (SIGMA) for fusion.

10. Cryopreserving solution: 10 % hybridoma grade dimethyl sulfoxide—DMSO (SIGMA) in FCS.
11. Reagents for ELISA.
    (a) ELISA plate reader.
    (b) ELISA 96-well plate.
    (c) Phosphate buffered saline (PBS): 137 mM NaCl, 2.7 mM KCl, 8 mM Na_2HPO_4, 2 mM KH_2PO_4.
    (d) Blocking buffer: 2 % bovine serum albumin (BSA) in PBS.
    (e) Washing buffer: PBS, pH 7.4, 0.05 % Tween 20.
    (f) Anti-mouse HRPO (SIGMA).
    (g) Substrate: *o*-phenylenediamine dihydrochloride (OPD, SIGMA).
    (h) Stop solution: H_2SO_4, 2 N.
    (i) sTNFR (80 kDa) human ELISA Module Set (Bender MedSystems).
12. For specific isotyping of mAbs use an antibody-capturing strips based kit (ROCHE).
13. The Affi-Gel protein A MAPS II (Monoclonal Antibody Purification System) kit (BIO-RAD).
14. Transfection with Lipofectamine reagent (Life Technologies).
15. Integra CELLine flasks for mAb concentrates (Integra Biosciences, Chur, Switzerland).
16. Standard laboratory equipment and supplies: sterile 96- and 24-well plates, tissue culture flasks (T25, T75, T175), plastic pipets (1, 5, 10, 25 mL), micropipets (20, 200, 1,000 μL), tips (200, 1,000 μL), pipettors, multichannel pipets, surgery dissection instruments, 1 mL 25GA sterile syringes (BD Plastipak, 300014), 15 and 50 mL conical tubes, 10 cm diameter culture petri dishes, hemocytometer, water baths, sterile tissue culture hoods, cell centrifuges, CO_2 humidified incubator, spectrophotometer, electrophoresis and transfer equipments, inverted and direct light microscope.

3 Methods

3.1 Monoclonal Antibody Production

3.1.1 Generation of Antigen

Alternative splicing can result in the generation of a protein containing a unique amino acid sequence due to changes in the open reading frame (Fig. 2). Order a 50 mg synthetic peptide of around 20 amino acids. Half of this peptide should be conjugated to keyhole limpet hemocyanin (KLH). The conjugated peptide will be used for the immunization and the unconjugated peptide for the screening of the mAbs. Several facilities and companies offer this service.

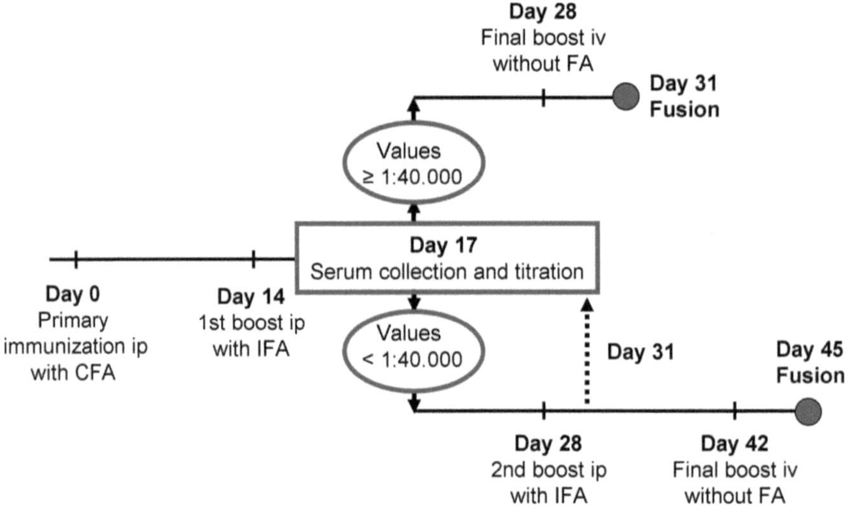

Fig. 3 Estimated timeline for the immunization protocol. *CFA* complete Freund's adjuvant, *IFA* incomplete Freund's adjuvant, *FA* Freund's adjuvant

The peptide has to be carefully selected in order to guarantee that is not present on other related family proteins or other splice variants.

3.1.2 Immunization

1. Immunize BALB/c mice (two to three mice) three to four times, at 2-week intervals, with the peptide of choice conjugated to KLH.
2. The first intraperitoneal (i.p.) injection consists of 40 μg of conjugated peptide in 200 μL of PBS mixed with 200 μL of complete Freund's adjuvant.
3. A second i.p. injection consists of 40 μg of conjugated peptide in 200 μL of PBS and 200 μL of incomplete Freund's adjuvant. This mixture is injected 2 weeks later in order to boost the initial immune response.
4. Collect the blood for serum titration the mice's tail 3 days after the second immunization. Determine the timeline for fusion by the values obtained in the serum titration (*see* Fig. 3). Therefore, titration values lower than 1/40,000 require a new immunization with antigen mixed with IFA (repetition of **steps 3** and **4**). In contrast, for values equal or higher than 1/40,000, only a last immunization is required (*see* **step 6**).
5. Perform the serum titration curve by ELISA, using unconjugated peptide-bound 96-well microplates (3 μg/mL). Add a serial serum dilution ranging from 1/1,000 to 1/160,000 to the ELISA plates. An anti-mouse HRPO is used as a detection antibody (see detailed procedure for this ELISA in Subheading 3.1.4). We measured the serum titre as the most

diluted sample with a positive OD signal (OD 2.5× higher than background noise).

6. The final injection consists of 40 μg of conjugated peptide/100 μL of PBS without Freund's adjuvant injected intravenously (i.v.).
7. Euthanize the mice on the third day after the final boost. Collect the blood and harvest the splenocytes.

3.1.3 Cell Fusion

1. NS1 myeloma cells should be thawed 1 week in advance and expand in flasks at a cell concentration $<10^6$ cells/mL in complete medium.
2. Split the NS1 cells the day before the fusion to guarantee their exponential growth. Usually, one T175 flask with a final volume of 100 mL is sufficient for a fusion.
3. Harvest NS1 cells in 50 mL conical tubes immediately before fusion. Spin down cells at room temperature (RT) and wash twice with RPMI without FCS also at RT. It is crucial that NS1 cells remain highly viable and free of protein in the medium.
4. Perform the cell fusion at 37 °C, thus it is important that all the reagents used are prewarmed.
5. Place a 15 mL conical tube with 1.5 mL of PEG solution in a water bath at 37 °C until fusion.
6. Euthanize the mouse. Collect blood from heart of the mouse, prepare serum, and keep it at −20 °C for further analysis. Cut abdominal skin and with the help of forceps expose the peritoneal cavity without damaging the peritoneal membrane. Next, cut the membrane with the help of sterile scissors and with a new set of sterile forceps extract the spleen and place it in a culture Petri dish with 10 mL of RMPI medium at RT (aseptic conditions are key to avoid contamination).
7. Isolate splenocytes from the spleen. Take two 1 mL syringes and bend needles 90°. Hold spleen with one of the syringes and with the help of the second syringe break the spleen's capsule. Carefully, scrape off splenocytes from spleen with the second syringe.
8. Transfer the splenocytes suspension to a 15 mL conical tube. Gently pipet up and down the cell suspension with a 5 mL pipet to break down aggregates. Let the debris to settle down for 5 min and then transfer cell suspension to a new 15 mL tube (carefully avoid taking any of the debris from the bottom). Wash cells twice with RPMI medium without serum and let splenocytes in a final volume of 10 mL of prewarmed RPMI at 37 °C.
9. Combine NS1 and splenocytes in a 50 mL conical tube. The number of obtained splenocytes varies depending on the immune response mounted against the antigen (usually around

120×10^6 splenocytes are obtained). We mix splenocytes with NS1 at ratio 4:1. Spin down at $200 \times g$ for 10 min at RT. Aspire supernatant. Dissociate the cell pellet by gently tapping on the bottom of the 50 mL tube.

10. Place 50 mL tube floating on the 250 mL beaker with 200 mL of water prewarmed at 37 °C. Add drop by drop 1 mL of warm PEG solution to the cell pellet using a 1 mL pipet, while constantly swirling the cell suspension to mix the cell pellet with PEG. The PEG addition should take approximately 1 min. Next, swirl cell suspension for one more minute.
11. At this point, start the dilution of PEG by adding first 1 mL of warm RPMI dropwise for 1 min (use 1 mL pipet). Then, add 5 mL of warm RPMI also dropwise for 1 min using a 5 mL pipet. Next, add 10 mL of RPMI for 1 min using a 10 mL pipet. Finally, add 20 mL of warm RPMI for 1 min using a 25 mL pipet.
12. Spin down cells at $200 \times g$ for 10 min at RT. Aspirate supernatant. Carefully resuspend the cell pellet by first slowly tapping the bottom of the 50 mL tube and then by adding 50 mL of warm HAT medium containing 20 % FCS and the growth supplement and slowly pipet up and down using a 25 mL pipet. Transfer cells to a bottle containing 270 mL of prewarmed HAT containing 20 % FCS and the growth supplement.
13. Transfer cells into sixteen 96-well-plates flat bottom, 200 μL per well, using a multichannel pipet with 200 μL tips (gently stir the bottle from time to time to ensure even distribution of cell suspension between plates).
14. Incubate at 37 °C in 5 % CO_2 in a humidified incubator. Importantly, this protocol does not require medium changes.

3.1.4 Screening

1. Start screening 10 days after the fusion. Under HAT selection the vast majority of cells die during the first 4 days. Usually, about 50 % of wells show hybridoma clones growing. In some wells more than one clone can be observed. The supernatant can be screened when the clone reached a 1/8 of the well area.
2. Perform three more screenings at 2-days intervals (a range of 400–800 clones are tested for each fusion).
3. From chosen wells, transfer 40 μL of hybridoma supernatant to a new prelabeled 96-well plate.
4. Dilute the hybridoma supernatant with 120 μL of PBS-BSA 2 % (dilution 4:1).
5. Perform the ELISA using unconjugated peptide-bound 96-well microplates. Plates are coated with 100 μL of 3 μg/mL of peptide in PBS overnight at 4 °C. Wash twice with 200 μL of washing buffer and block with 250 μL of blocking buffer at

37 °C for 1 h. After washes (3×) with washing buffer, 100 μL of diluted hybridoma supernatant is added per well. After 1 h incubation, plates are washed extensively (4×) with washing buffer. Next, incubate plates with 100 μL of an anti-mouse HRPO antibody (0.5–1 μg/mL) for 30 min. Wash twice with washing buffer and twice with PBS. Develop with 100 μL of OPD, incubating for 15 min at RT. Add 50 μL per well of stop solution and measure OD with ELISA reader (405 nm). Supernatants that present an OD 2.5× over the background are considered positive. As a positive control mouse serum from the immunized mouse is used.

6. Transfer the hybridomas of positive wells to 24-well plates (1.5 mL/well) and grow until they are confluent in medium supplemented with HT instead of HAT. Then, perform a second screening in order to test if the hybridoma still produces the desired antibody. At this point the specificity of the antibodies should be tested with other techniques such as western blot, immunoprecipitation, or reactivity with transfected cells (*see* **Note 2**). Selected clones are grown in T75 flasks and the cells are frozen with 10 % DMSO. Freezing at least two vials of each selected hybridoma is recommendable at this point.

3.1.5 Cloning by Limiting Dilution

1. The number of selected clones varies between fusions. An optimal number of positive hybridomas to be subcloned range between 2 and 4 per fusion. This step is essential to guaranty that the antibody will be produced by a single clone.
2. Pipet out 150 μL of cells from the desired 96-well by mixing up and down with an automatic pipet with 200 mL tips. Transfer this volume to 1 mL of HT medium and count cells. Transfer the volume corresponding 2×10^3 cells into 40 mL of medium placed in a 50 mL conical tube (*dilution A*—10 cells/well). Next, transfer 20 mL from dilution A to the next tube containing 20 mL of HT (*dilution B*—5 cells/well). Repeat this procedure twice to achieve *dilutions C* and *D* (2.5 and 1.25 cells/well respectively).
3. Pipet out 200 μL from dilution D into a 96-well plate using a multichannel pipet (one 96-well plate per dilution). Repeat same protocol for each dilution.
4. Incubate at 37 °C in 5 % CO_2 in a humidified incubator for 10–15 days.
5. Retest 20 single clones by ELISA with the unconjugated peptide from the lowest dilutions plates (usually D or C). Repeat the cloning protocol with one of the positive clones.
6. Every selected hybridoma clone may require two more consecutive rounds of recloning. This protocol should be repeated until all 20 selected subclones are positive for the screening procedure of choice.

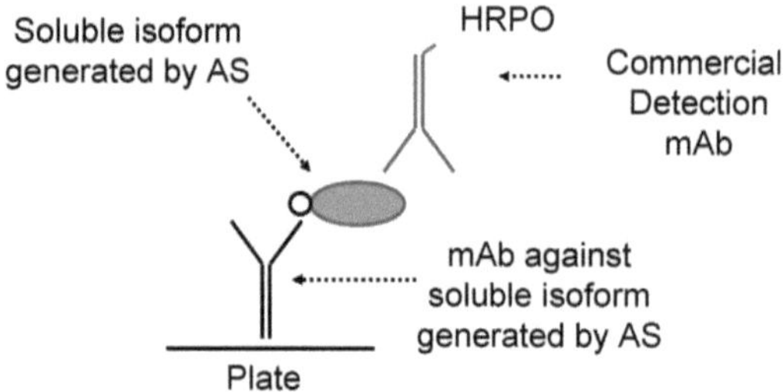

Fig. 4 Sandwich ELISA for the detection of soluble receptor generated by alternative splicing (AS)

7. Since isotype is one of the features that will affect mAb performance in the application of choice, at this point, we also determine the isotype of the mAb (class and subclass). We use an IsoStrip mouse monoclonal antibody Isotyping Kit to determine the isotype, although multiple kits are commercially available.

3.1.6 Expansion of Hybridoma Clones, Production of Concentrate Monoclonal Antibody Supernatant, and Protein Purification

1. Expand the selected hybridoma cells, after cloning by limited dilution in a T75 flask with RMPI with 15 % of FCS. (No hybridoma fusion and cloning supplement are added.)
2. Mix eight vials containing $10–20 \times 10^6$ cells with cryopreserving solution and cryopreserve.
3. Next, culture the hybridoma cells in CELLine CL350 or CL1000 flasks following manufacture's indications.
4. Concentrate mAb supernatants collected from CELLine CL350 or CL1000 flasks, filter and save at −20 °C until purification.
5. Purify the mAbs with a Protein A column following manufacture's indications.
6. Finally, dialyze the purified mAb extensively against sterile PBS utilizing dialysis cassette. Measure the protein quantification by spectrophotometry and check the purity of the protein by SDS-polyacrylamide gel electrophoresis. Aliquot mAb and store at −20 °C.

3.2 Production of a Sandwich ELISA for the Detection of Soluble Receptor Generated by Alternative Splicing

The newly generated mAbs are tested for their capacity to function as capturing antibodies in a sandwich ELISA for the specific quantification of the cytokine soluble protein generated by alternative splicing (*see* Fig. 4).

3.2.1 ELISA Detection of Soluble Isoform Generated by Alternative Splicing in Supernatants of COS-Transfected Cells (See **Note 3**)

1. Generate the supernatants containing soluble receptor form COS transfected cells. COS-7 cells are transiently transfected with Lipofectamine reagent following manufacturer's indications with empty vector (negative transfection control) or with vector-cDNA full-length receptor or cloned vector with cDNA encoding for soluble isoform generated by AS cDNA. Collect the supernatants 48 h after transfection.

2. Coat the ELISA microplates with 100 μL (3 μg/mL) of mAbs directed against soluble isoform generated by alternative splicing, leave overnight at 4 °C, and block with blocking buffer.
3. Wash 3× with 200 μL of washing buffer.
4. Incubate for 1 h with 100 μL of COS transfected supernatants diluted 1:10 or 1:5 in blocking buffer.
5. As detection antibody, use a commercial mAb peroxidase-conjugated which recognizes the receptor's ectodomain. Incubate for 60 min.
6. Wash plates twice with wash buffer and twice with PBS. Develop with 100 μL of OPD by incubating for 15 min at RT. Add stop solution and measure OD with ELISA reader (405 nm) as described in **step 5** in Subheading 3.1.4.

The capturing antibody giving the highest signal:noise ratio is chosen for the development of the ELISA to test patient samples. The unwanted cross reactivity with shed isoforms should be tested. COS-7 transfected with the vector-cDNA full-length receptor are treated with the shedding inducing agent PMA (50 ng/mL). The ELISA for the detection of soluble receptor generated by alternative splicing (AS) shouldn't recognize the receptor generated by shedding.

3.2.2 ELISA Detection of Soluble Receptor Generated by Alternative Splicing in Patients Sera

1. Identical to the protocol in Subheading 3.2.1, here we use a modified version of the commercial ELISA kit (sTNFR (80 kDa) human ELISA Module Set) in order to quantify the soluble isoform generated by alternative splicing in the patients sera. Thus, the newly generated mAb is used as a capture antibody, whereas the peroxidase-conjugated anti-soluble receptor provided by the commercial kit is used as a detection antibody. As a standard, we use a pool of supernatants generated in COS transfected cells, containing a known concentration of the recombinant soluble receptor.
2. Dilute the sera 1:10 or 1:5 in blocking buffer. The sensitivity of this ELISA should be identical to that of the commercial ELISAs for the detection of the soluble receptors with the lower detection levels around 0.5–1 ng/mL [12].

4 Notes

1. Selecting the right FCS for the fusions requires testing several batches of FCS for the optimal growth of your hybridomas. The best method to test the quality of FCS is by using them in limiting dilution assays. RMPI medium with 10 % of preselected FCS batch plus supplements is used to culture NS1 cell

lines. For hybridomas growth we use a RMPI medium with 20 % FCS plus supplements (4 mM L-glutamine, 50 U penicillin, 50 μg streptomycin).

2. Strict validation of newly generated mAbs is critical. In this chapter, we describe a first validation of mAbs for their application in ELISA. However, further validation by other methods is desirable. We recommend to validate the antibodies using cell staining and immunochemical methods.

   For antibody validation by immunofluorescence, COS-7 cells are transfected with FL and AS cDNAs. The cells are placed on sterile glass coverslips in 24-well culture plates. After 48 h, transfected cells with AS cDNA are treated with 1 mg/mL of the protein transport inhibitor Brefeldin A for 4 h at 37 °C, 5 % CO_2 in order to retain the soluble protein in the cytoplasm. Immunocytochemistry analysis is performed using indirect immunofluorescence. Briefly, fix and permeabilize the cells using 1.5 % formaldehyde and 0.1 % Triton. Then wash the cells once with PBS containing 2 % FCS. Incubate the cells with PBS containing 10 % FCS for 30 min and then stain with the newly generated mAb for 30 min at room temperature (1/4 diluted supernatant with blocking buffer). Wash the cells 3× with PBS containing 2 % FCS for 5 min, incubate with anti-mouse fluorochrome-conjugated antibody for 30 min at RT. Wash cell and fix in methanol (−20 °C) for 15 min. Mount coverslips on slides with mounting medium and examine samples by fluorescence microscopy. A commercial antibody against the receptor ectodomain can be used as a positive control. Purified mouse IgG can be used as a negative control. Immunocytochemical analysis utilizing the antibody against the receptor ectodomain should reveal an intense staining of both the cell surface and cytoplasm of cells transfected with the full-length receptor, but no significant cell surface staining should be detected in COS-7 cells transfected with the soluble isoform.

   For antibody validation by immunoprecipitation and immunoblotting, collect the supernatant after 72 h of culture of transfected COS-7 cells with FL and AS cDNAs. Immunoprecipitations and western blots are carried out using standard protocols of 5× concentrated supernatants.

3. A previous step required for the COS-7 transfection mentioned in Subheading 3.2 is the cloning of full-length (FL) receptor and isoform generated by alternative splicing (AS) in expression vectors. Briefly, isolate total RNA from human cells expressing full-length and soluble isoform receptor generated by AS with TRIzol reagent (Life Technologies). Synthesize cDNA from 1 μg of total RNA using the First-Strand cDNA synthesis kit (Boehringer Mannheim, Mannheim, Germany).

A PCR product generated with the help of primers designed to specifically amplify the FL receptor or primers to amplify the soluble form generated by AS. PCR products are cloned in expression vector. Sequencing is performed with the Big Dye terminator kit (Perkin-Elmer, Foster City, CA), according to the manufacturer's instructions. Alternatively several companies have these cDNAs available in their catalogues.

References

1. Rose-John S, Heinrich PC (1994) Soluble receptors for cytokines and growth factors: generation and biological function. Biochem J 300:281–290
2. Heaney ML, Golde DW (1996) Soluble cytokine receptors. Blood 87:847–857
3. Levine SJ (2008) Molecular mechanisms of soluble cytokine receptor generation. J Biol Chem 283:14177–14181
4. Heaney ML, Golde DW (1998) Soluble receptors in human disease. J Leukoc Biol 64: 135–146
5. Ernandez T, Mayadas TN (2009) Immunoregulatory role of TNFalpha in inflammatory kidney diseases. Kidney Int 76:262–276
6. Heemann C, Kreuz M, Stoller I, Schoof N, von Bonin F, Ziepert M et al (2012) Circulating levels of TNF receptor II are prognostic for patients with peripheral T-cell non-Hodgkin lymphoma. Clin Cancer Res 18:3637–3647
7. Hivert MF, Sullivan LM, Shrader P, Fox CS, Nathan DM, D'Agostino RB Sr et al (2010) The association of tumor necrosis factor alpha receptor 2 and tumor necrosis factor alpha with insulin resistance and the influence of adipose tissue biomarkers in humans. Metabolism 59:540–546
8. Gohda T, Niewczas MA, Ficociello LH, Walker WH, Skupien J, Rosetti F et al (2012) Circulating TNF receptors 1 and 2 predict stage 3 CKD in type 1 diabetes. J Am Soc Nephrol 23:516–524
9. Aderka D, Wysenbeek A, Engelmann H, Cope AP, Brennan F, Molad Y et al (1993) Correlation between serum levels of soluble tumor necrosis factor receptor and disease activity in systemic lupus erythematosus. Arthritis Rheum 36:1111–1120
10. Gabay C, Cakir N, Moral F, Roux-Lombard P, Meyer O, Dayer JM et al (1997) Circulating levels of tumor necrosis factor soluble receptors in systemic lupus erythematosus are significantly higher than in other rheumatic diseases and correlate with disease activity. J Rheumatol 124:303–308
11. Schröder J, Stüber F, Gallati H, Schade FU, Kremer B (1995) Pattern of soluble TNF receptors I and II in sepsis. Infection 23:143–148
12. Lainez B, Fernandez-Real JM, Romero X, Esplugues E, Cañete JD, Ricart W et al (2004) Identification and characterization of a novel spliced variant that encodes human soluble tumor necrosis factor receptor 2. Int Immunol 16:169–177
13. Cañete JD, Albaladejo C, Hernández MV, Laínez B, Pinto JA, Ramírez J et al (2012) Clinical significance of high levels of soluble tumour necrosis factor-α receptor-2 produced by alternative splicing in rheumatoid arthritis: a longitudinal prospective cohort study. Rheumatology (Oxford) 50:721–728

Chapter 17

Effective Expression and Purification of Bioactive Recombinant Soluble LIGHT

Isamu Tsuji, Keiji Iwamoto, and Yasushi Shintani

Abstract

LIGHT (TNFSF14, CD258) is a member of the tumor necrosis factor (TNF) ligand superfamily, which is involved in innate and adaptive immune responses as well as in regulation of cell survival and proliferation. LIGHT forms a membrane-anchored homotrimeric complex on the cell surface and is often processed as a soluble protein. In this study, we established an effective strategy for producing bioactive soluble forms of human LIGHT (s-hLIGHT), which is an extracellular region (Ile^{84}-Val^{240}) of human LIGHT (hLIGHT), and soluble forms of mouse LIGHT (s-mLIGHT), which is an extracellular region (Asp^{72}-Val^{239}) of mouse LIGHT (mLIGHT). To enhance the refolding of s-hLIGHT from inclusion bodies in *Escherichia coli*, we added L-cysteine to the denaturation buffer, which significantly improved the refolding efficiency of s-hLIGHT. However, there was little information available about the biological activity of mLIGHT in the literature because of the difficulty in producing bioactive s-mLIGHT. We produced trimeric s-mLIGHT by fusing s-mLIGHT with a trimerization domain termed "foldon" from bacteriophage T4 fibritin (Foldon-mLIGHT). We further demonstrated that Foldon-mLIGHT inhibited the growth of mouse carcinoma cells at the picomolar range, indicating that trimerization of s-mLIGHT is essential for its biological activity.

Key words LIGHT, *Escherichia coli*, Inclusion body, 293F, L-cysteine, Foldon

1 Introduction

LIGHT is a member of the tumor necrosis factor (TNF) ligand superfamily, which is expressed in activated T-cells and immature dendritic cells (DCs) [1, 2]. LIGHT enhances T-cell stimulation and induces IL-12 secretion and DC maturation by upregulating the expression of co-stimulatory molecules [3]. Constitutive expression of LIGHT results in T-cell-mediated inflammation and tissue destruction [4], whereas the absence of LIGHT impairs T-cell responses [2]. LIGHT binds to the lymphotoxin β receptor (LTβR) and herpes virus entry mediator (HVEM), which are TNF receptor superfamily members that have distinct expression patterns [1, 2]. Notably, the decoy soluble receptor TR6/DcR3 is known to inhibit LIGHT activities [5]. The amino acid sequence of

Jagadeesh Bayry (ed.), *The TNF Superfamily: Methods and Protocols*, Methods in Molecular Biology, vol. 1155, DOI 10.1007/978-1-4939-0669-7_17, © Springer Science+Business Media New York 2014

LIGHT is approximately 30 % homologous to those of Fas ligand, TNFβ, and LTβ. Human LIGHT (hLIGHT) consists of 240 amino acid residues, a 37 amino acid N-terminal intracellular region, a transmembrane domain consisting of 22 amino acids, and a C-terminal extracellular region of 181 amino acids. The region between amino acids 90 and 240 is considered the receptor-binding region, with a sequence similarity of approximately 20–30 % with other TNF superfamily proteins [6]. Similar to TNFα, LIGHT is thought to be expressed as a trimer on the cell surface and is converted into a soluble form by proteolytic cleavage [7]. Gel filtration analysis has shown that the extracellular domain of recombinant hLIGHT forms a trimer. The substitution of Gly119, Tyr173, and Arg228, which are located far apart in the monomeric structure of hLIGHT, reduces its binding affinity to the receptor; hence, hLIGHT is thought to bind to its receptor as a trimer [8, 9]. On the other hand, mLIGHT is 239 amino acids long, and the mouse and human LIGHT extracellular regions have a 74 % amino acid sequence similarity.

The biological activities of hLIGHT have been investigated by transient expression of s-hLIGHT in insect Sf9 [10, 11] and COS cells [12]. Although these expression systems are convenient for obtaining bioactive proteins quickly, they are not suitable for large-scale production for in vivo evaluation and clinical trials because of their inefficiency and high cost. Recombinant proteins expressed as inclusion bodies in *Escherichia coli* are widely used for the large-scale production of therapeutic proteins that do not require post-translational modifications. The high expression yield and simple recovery of inclusion bodies from host cells are industrially attractive, although denaturation and refolding from inclusion bodies are required to obtain bioactive protein.

Our previous study showed that the addition of L-cysteine to the denaturation buffer significantly improves the refolding efficiency of s-hLIGHT by reverting the denatured protein into its native conformation at a high yield [13]. The effect of L-cysteine was synergistically enhanced by addition of L-arginine in the refolding buffer. The optimal concentrations of L-cysteine and L-arginine in the denaturation and refolding buffers were 8 mM and 0.8 M, respectively. However, unlike hLIGHT, there are only a few reports in the literature about the production of bioactive recombinant mLIGHT because of the difficulty in producing bioactive s-mLIGHT [14]. Here, we show a method for producing bioactive s-mLIGHT (Foldon-mLIGHT) by using a 27 amino acid peptide called foldon, the C-terminal domain of bacteriophage T4 fibritin [15]. Foldon-mLIGHT was secreted from 293F cells as a 68-kDa trimeric protein, and it potently inhibited the growth of the FM3A mouse mammary carcinoma cell line with an IC_{50} of 77 pM, although mLIGHT monomers or dimers showed weak or no inhibitory activity. Flow cytometry analysis revealed that

Foldon-mLIGHT also bound specifically to HVEM-transfected cells [16]. In this chapter, we describe the production methodologies for recombinant s-hLIGHT and Foldon-mLIGHT from *E. coli* and 293F cells, respectively.

2 Materials

2.1 Effective Production of Bioactive s-hLIGHT by Escherichia coli

2.1.1 Expression

1. Modified M9 medium: 16.8 g/L Na_2HPO_4, 3.0 g/L $KHPO_4$, 1.0 g NH_4Cl, 0.5 g/L NaCl, 15 g/L glucose, 0.246 g/L $MgSO_4$, 10 g/L HyCase amino (Sigma Aldrich), 15 g/L Bacto-yeast extract (BD Biosciences).
2. *Escherichia coli* (*E. coli*) MM294 (DE3) competent cells treated with 0.1 M calcium chloride (1×10^9 cells/tube).
3. Isopropyl β-thiogalactoside (IPTG) (Wako Pure Chemicals, Japan).
4. Luria-Bertani's medium: 10 g/L tryptone, 5 g/L yeast extract, 10 g/L NaCl (Wako Pure Chemical, Japan).
5. Restriction enzymes: *Nde*I, *Bam*HI.

2.1.2 Isolation and Purification

1. Extraction buffer: 50 mM Tris–HCl, pH 7.4, 2 mM $CaCl_2$.
2. Pellet wash buffer: 50 mM Tris–HCl, pH 7.4, 2 M guanidine hydrochloride (GuHCl), 2 mM $CaCl_2$.
3. Denaturation buffer: 100 mM Tris–HCl, pH 7.4, 3.5 M GuHCl, 2 mM $CaCl_2$, 8 mM L-cysteine.
4. Refolding buffer: 50 mM Tris–HCl, pH 8.0, 0.8 M L-arginine.
5. POROS® HS 50 μm column (column size: 130 mL packed in house, Applied Biosystems) with Gilson Model 305 HPLC system (200 WTi pump head, flow rate: 50 mL/min, detection: 280 nm).
6. POROS® HS 50 μm binding buffer: 50 mM sodium acetate buffer, pH 5.8.
7. POROS® HS 50 μm elution buffer: 50 mM sodium acetate buffer, pH 5.8, 1 M NaCl.
8. TSK-Gel CM-5PW column (column size: 21.5 × 150 mm, Tosoh Corporation, Japan) with Gilson Model 305 HPLC system (10 WTi pump head, flow rate: 5 mL/min, detection: 280 nm).
9. 1st TSK-Gel CM-5PW binding buffer: 50 mM MES-NaOH, pH 6.6, 0.15 M NaCl.
10. 1st TSK-Gel CM-5PW elution buffer: 50 mM Tris–HCl, pH 8.0, 0.5 M NaCl.

11. 2nd TSK-Gel CM-5PW binding buffer: 50 mM sodium acetate buffer, pH 6.0, 0.15 M NaCl.
12. 2nd TSK-Gel CM-5PW elution buffer: 50 mM sodium acetate buffer, pH 6.0, 1.5 M NaCl.
13. Asahipak C4P-50 (column format: 4.6 mmD × 250 mmL, Showa Denko, Japan).
14. Asahipak C4P-50 analysis buffer A: 0.1 % trifluoroacetic acid (TFA), 27 % acetonitrile.
15. Asahipak C4P-50 analysis buffer B: 0.1 % TFA, 80 % acetonitrile.

2.1.3 Cytotoxicity Assay

1. WiDr, a human colon adenocarcinoma cell line (ATCC® number: CCL-218™).
2. 96-well cell culture plates.
3. 0.25 % Trypsin-EDTA.
4. Dulbecco's phosphate buffered saline (D-PBS).
5. Eagle's minimum essential medium (EMEM).
6. Nonessential amino acids.
7. Fetal bovine serum (FBS).
8. Human Interferon-gamma (IFN-γ, R&D Systems).
9. 5-Bromo-2′ deoxyuridine (BrdU) ELISA kit (Roche Diagnostics).

2.1.4 Equipments

1. Bioreactor for culturing *E. coli* (20-L scale).
2. Floor-type centrifuge (Avanti HP-20; Beckman/coulter).
3. Sonifier 450 Cell Disruptor with 3/4" high gain horn, 101-147-035 (Branson Ultrasonics, Danbury, CT).
4. Laboratory stirrers.
5. Ultrafiltration devices (molecular weight cutoff: 10 kDa, 0.1 m^2 Hydrosart®) (Sartorius, Germany).
6. HPLC System: Model 305 HPLC system (Gilson).
7. Ultrafiltration centrifuge (Centricon YM-10, Millipore).
8. SDS-PAGE equipment (Bio-Rad, Daiichi Pure Chemicals).
9. Microplate reader (SpectraMax 340PC, Molecular Devices).

2.2 Effective Production of Bioactive Soluble Mouse LIGHT, Foldon-mLIGHT, by FreeStyle™ 293-F Cells

2.2.1 Expression

1. Superscript mouse 8.5-day embryo cDNA library (Invitrogen).
2. Oligo DNA primers (Invitrogen).
3. Big Dye terminator cycle sequencing kit (Applied Biosystems).
4. pFLAG-CMV3 vector (Sigma).
5. Takara Pyrobest DNA polymerase (Takara).
6. Restriction enzymes.
7. Plasmid purification kit.

2.2.2 Isolation and Purification

1. FreeStyle™ 293-F cells, adapted to serum-free suspension culture (Invitrogen).
2. FreeStyle™ 293-F expression medium (Invitrogen).
3. Opti-MEM I (Invitrogen).
4. 293fectin™ reagent (Invitrogen).
5. Anti-Flag mAb M2 agarose (Sigma).
6. Econo-column (Bio-Rad).
7. ActiClean Etox column (Sterogene).
8. Phosphate buffer saline (PBS), pH 7.4.
9. 0.5 M arginine-HCl, pH 3.8.
10. Superdex 200 16/60 column (GE Healthcare).
11. 10 mM sodium phosphate, pH 7.4, 1 M NaCl.
12. 10–20 % SDS-polyacrylamide gels.
13. SDS-PAGE Standards (Bio-Rad).
14. SimplyBlue™ SafeStain (Invitrogen).
15. Superdex 200 10/300 GL column (GE Healthcare).
16. Gel filtration standard (Bio-Rad).

2.2.3 Cytotoxicity Assay

1. Mouse mammary carcinoma cell line FM3A (European Collection of Cell Cultures).
2. RPMI 1640 medium.
3. Fetal bovine serum (FBS).
4. Mouse IFN-γ (PeproTech).
5. Cell Counting Kit-8 (Dojindo).
6. Mouse LTβR-Fc (R&D systems).
7. Mouse IgG isotype control (R&D systems).

2.2.4 Equipments

1. GeneAmp® PCR system (Applied Biosystems).
2. Genetic analyzer (Applied Biosystems).
3. Orbital shaker.
4. CO_2 incubator.
5. 1-L polycarbonate, disposable, sterile Erlenmeyer flasks.
6. Refrigerated centrifuge (Beckman Coulter).
7. Liquid chromatography system (GE Healthcare).
8. Polyacrylamide gel electrophoresis (PAGE) equipment.
9. Microplate reader SPECTRA Max 340PC (Molecular Devices).

3 Methods

3.1 Production and Characterization of Bioactive s-hLIGHT by E. coli

3.1.1 Transformation of E. coli with s-hLIGHT Plasmid

1. To construct an s-hLIGHT plasmid, amplify the cDNA coding the extracellular region (Ile^{84}-Val^{240}) of hLIGHT from a full-length LIGHT cDNA by PCR using 5′-TATACATATG ATACAAGAGCGAAGGTC-3′ as the forward primer and 5′-AGCCGGATCCGACCTCACACCATGAAA-3′ as the reverse primer.
2. Digest the PCR product with *Nde*I and *Bam*HI, and ligate it into the *Nde*I-*Bam*HI site of the pTCII expression vector.
3. Introduce the plasmid into *E. coli* MM294 (DE3) competent cells, and screen transformants expressing s-hLIGHT by SDS-PAGE.

3.1.2 Cell Culturing

1. Inoculate transformant cells into 500 mL of Luria–Bertani's medium in a 2-L Erlenmeyer shaker flask, and culture the flask overnight in a 30 °C incubator on a 5.1 cm orbital shaker rotating at 200 rpm.
2. Transfer the culture broth into 18 L modified M9 medium, and culture the transformant at 37 °C in a 20-L bioreactor. During cultivation, maintain the pH of the culture medium at 6.8, keep the dissolved oxygen concentration at 2 ppm, and feed L-GLUCOSE into the bioreactor.
3. Induce s-hLIGHT expression by adding IPTG into the culture medium at a final concentration of 0.15 mM, when a cell density of 8 (OD_{600}) is reached.
4. After incubation for 3 h, harvest the cells by centrifugation at 2,500 × *g* for 10 min. The pellets should then be sonicated, and the soluble and insoluble fractions should be analyzed by SDS-PAGE.

3.1.3 Isolation of Inclusion Bodies and Renaturation of s-hLIGHT

To obtain highly purified s-hLIGHT, inclusion bodies should be isolated with low protein contamination from *E. coli* cells by repeated sonication. Following this, the inclusion bodies should be solubilized in a denaturation buffer containing L-cysteine and rapidly diluted in the refolding buffer containing L-arginine. The combination of L-cysteine in the denaturation buffer and L-arginine in the refolding buffer is essential to obtain a high yield of s-hLIGHT. All steps should be performed in the cold room as follows:

1. Suspend 200 g of the frozen cell pellet of s-hLIGHT-expressing *E. coli* in 1 L of extraction buffer and disrupt it by ultrasonication for 60 min.
2. Centrifuge the lysate at 11,300 × *g* for 60 min and resuspend the pellet in 500 mL of the same buffer.

3. Disrupt the pellet again by ultrasonication for 20 min (*see* **Note 1**).
4. After centrifugation at 11,300×*g* for 60 min, resuspend the pellet in 500 mL of pellet-wash buffer.
5. Disrupt the lysate again by ultrasonication for 20 min (*see* **Note 2**).
6. After centrifugation at 11,300×*g* for 60 min, solubilize the pellet in 600 mL of denaturation buffer with overnight stirring using a laboratory stirrer (*see* **Note 3**).
7. After centrifugation at 11,300×*g* for 60 min on the second day, refold the denatured proteins in the supernatant by rapid dilution with 12 L of refolding buffer and let it stand for 17 h at 4 °C.
8. Concentrate the refolded LIGHT solution up to a volume of 250 mL using a 0.1 m^2 Hydrosate® ultrafiltration cassette and add 500 mL of POROS® HS binding buffer to the concentrated solution (*see* **Note 4**).
9. After centrifugation at 11,300×*g* for 20 min, dilute the crude s-hLIGHT solution with 2 L distilled water, and immediately apply it to a POROS® HS chromatography column (*see* **Note 5**).

3.1.4 Purification of s-hLIGHT

As s-hLIGHT has no purification tag, cation-exchange column chromatography should be performed for the s-hLIGHT purification. The purity of s-hLIGHT can be determined by SDS-PAGE or reverse phase columns as described in Subheading 3.1.5.

1. Use a POROS® HS column for the first purification step. The column should be equilibrated with POROS® HS 50 μm binding buffer.
2. After application of s-hLIGHT solution to the column, wash the column using the same buffer, with at least ten times the column volume.
3. Elute the bound protein with a 30 min linear gradient of NaCl that ranges in concentration from 0 to 0.6 M, by mixing POROS® HS 50 μm elution buffer with POROS® HS 50 μm binding buffer (*see* **Note 6**). Before collection, s-hLIGHT proteins in the eluted solution should be determined by SDS-PAGE or reversed-phase HPLC.
4. Use a TSK-GEL CM-5PW column for the next purification step. The column should be equilibrated with 1st TSK-Gel CM-5PW binding buffer.
5. Dilute one part of s-hLIGHT-containing solution with two parts of distilled water and apply it to the column (*see* **Note 7**).
6. After application of the s-hLIGHT solution, wash the column with at least twice the column volume of the same buffer, and

elute the bound proteins with 50-min linear pH gradient, from 6.6 to pH 8.0, by mixing 1st TSK-Gel CM-5PW elution buffer with 1st TSK-Gel CM-5PW binding buffer (*see* **Note 8**). As performed previously, s-hLIGHT proteins in the eluted solution should be determined by SDS-PAGE or reversed-phase HPLC before collection.

7. For the final step, TSK-GEL CM-5PW chromatography should be performed again using the same conditions. Briefly, s-hLIGHT solution should be diluted (1:2) with distilled water before application to the column equilibrated with 2nd TSK-Gel CM-5PW binding buffer (*see* **Note** 7).
8. Wash the column with at least twice the column volume of the same buffer, and s-hLIGHT proteins should elute out with a 30-min linear gradient of NaCl, with the concentration ranging from 0.15 to 0.6 M, by mixing 2nd TSK-Gel CM-5PW elution buffer with 2nd TSK-Gel CM-5PW binding buffer (*see* **Note 9**). As performed previously, s-hLIGHT proteins in the eluted solution should be determined by SDS-PAGE or reversed-phase HPLC before collection.
9. Change the s-hLIGHT solution to 2nd TSK-Gel CM-5PW binding buffer before concentrating it to 1 mg/mL by using a Millipore Centricon YM-10.

3.1.5 Determination of s-hLIGHT Purity

The purity of s-hLIGHT can be determined by SDS-PAGE or reversed-phase column chromatography as follows:

1. Use an Asahipak C4P-50 column equipped with HPLC GILSON Model 305 system. Samples should be applied to the column equilibrated with Asahipak C4P-50 analysis buffer A. Elute at a flow rate of 0.5 mL/min with a 40-min linear gradient of acetonitrile ranging from 27 to 56 %, by mixing Asahi C4P-50 analysis buffer B with Asahi C4P-50 analysis buffer A. The eluate should be monitored using a UV detector (280 nm). The final s-hLIGHT sample elutes as a single peak with a purity of 99.8 % or higher.
2. Perform SDS-PAGE according to the method described by Laemmli [17] on a PAG mini 15/25 in the presence of 0.1 % SDS under reducing conditions. After electrophoresis, the gels should be stained using Quick CBB stain kit. Purified s-hLIGHT migrates as a single band on SDS-PAGE under the reduced condition with a purity of 99.8 % or higher.

3.1.6 Storage

s-hLIGHT should be dissolved in 2nd TSK-Gel CM-5PW binding and stored in stock tubes at −80 °C (*see* **Note 10**). The cytotoxicity of s-hLIGHT does not change for at least 1 year under these storage conditions.

3.1.7 s-hLIGHT Characterization by Cytotoxicity Assay

Cytotoxicity of s-hLIGHT can be measured by the proliferation assay using WiDr cells and BrdU ELISA kit as follows:

1. For maintenance of the human colon adenocarcinoma cell line WiDr, use EMEM supplemented with nonessential amino acids and 10 % FBS at 37 °C in a 5 % CO_2 atmosphere.
2. Seed WiDr cells at 5,000 cells/well in 96-well plates and culture in the presence of various concentrations of s-hLIGHT (up to 100 ng/mL) and IFN-γ (200 U/mL) for 3 days.
3. Treat the cells with BrdU in a CO_2 incubator and with peroxidase-labeled anti-BrdU for 1.5 h according to the instructions in the BrdU ELISA kit manual. The absorbance at 340 nm/492 nm is measured on a plate reader.

3.2 Production and Characterization of Foldon-mLIGHT, by FreeStyle™ 293-F Cells

3.2.1 Vector Construction

1. Superscript mouse 8.5-day embryo cDNA library can be used to obtain mLIGHT cDNA by using PCR with 5′-TCTGCTCTGGCATGGAGAGTGTGGT-3′ as the forward primer and 5′-CTATTGCTGGGTTTGAGGTGAGTC-3′ as the reverse primer (identical to GenBank NM_019418).
2. Amplify cDNA encoding extracellular domains of mLIGHT (Asp^{72}-Val^{239}) by PCR using 5′-GATGGAGGCAAAGGCT CCTGGG-3′ as the forward primer and 5′-ATGAATTCTCA GACCATGAAAGCTCCGAAA-3′ as the reverse primer.
3. Ligate synthetic oligonucleotides encoding a foldon peptide sequence, GYIPEAPRDGQAYVRKDGEWVLLSTFL (a 27 amino acid trimerization domain derived from the native T4 phage fibritin) to the N-terminus of an extracellular domain of s-mLIGHT to make a Foldon-mLIGHT cDNA with a flanking *Hind*III site at the N-terminus.
4. Digest the Foldon-mLIGHT cDNA with *Hind*III and *Eco*RI and ligate it into *Hind*III/*Eco*RI digested vector pFLAG-CMV3.
5. Introduce the plasmid into *E. coli* DH5α competent cells, and screen transformants containing the Foldon-mLIGHT expression vector by PCR.
6. Purify plasmid DNA for transformation with a Qiagen Plasmid Maxi kit.

3.2.2 Transformation with Foldon-mLIGHT Plasmid

1. Dilute 0.3 mg of the pFLAG-CMV3 expression vector encoding the Foldon-mLIGHT cDNA in Opti-MEM I to a total volume of 10 mL.
2. Dilute 0.6 mL of 293fectin in Opti-MEM I to a total volume of 10 mL, mix gently, and incubate for 5 min at room temperature.
3. Add the diluted DNA to the diluted 293fectin to obtain a total volume of 20 mL, and incubate for 20–30 min at room temperature to allow the DNA-293fectin complexes to form.

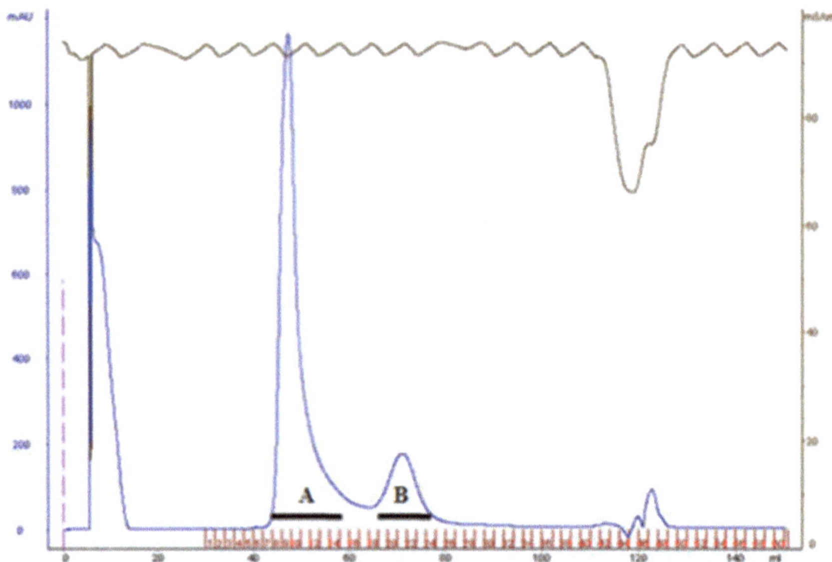

Fig. 1 Purification of Foldon-mLIGHT by size-exclusion chromatography. The *blue line* shows UV absorbance at 280 nm, and the *brown line* shows conductivity. Foldon-mLIGHT was separated as two peaks: *upper peak* (*A*) and *lower peak* (*B*)

4. Suspend 3×10^8 cells of FreeStyle™ 293-F cells to fresh 280 mL of FreeStyle™ 293-F expression medium in a 1-L Erlenmeyer shaker flask.
5. Add 20 mL of DNA-293fectin complex to the flask (final cell density of 1×10^6 cells/mL) and place the flask in a 37 °C incubator on a 5.1 cm orbital shaker rotating at 125 rpm with a humidified atmosphere of 8 % CO_2.
6. After 3 days of culture, collect the supernatant by centrifugation at $1,000 \times g$ for 10 min and filter with a 0.22-μm filter to remove any remaining cells.

3.2.3 Foldon-mLIGHT Isolation and Purification by Column Chromatography

Foldon-mLIGHT protein can be purified by affinity chromatography using 3-mL anti-Flag mAb M2 agarose column as follows:

1. Equilibrate an anti-Flag mAb M2 agarose with PBS buffer at five times the column volume, and load the sample onto the column under gravity flow.
2. Wash the column with 10 column volumes of PBS buffer, and elute the bound protein with 0.5 M arginine-HCl (pH 3.8).
3. Neutralize the pH by adding 1/10th the volume of 1 M Tris–HCl (pH 8.0). The eluted Foldon-mLIGHT proteins can be further purified by size-exclusion chromatography (SEC) using Superdex 200 16/60 column in PBS (pH 7.4) at a flow rate of 0.75 mL/min to remove its aggregates (*see* Fig. 1 and **Note 11**).

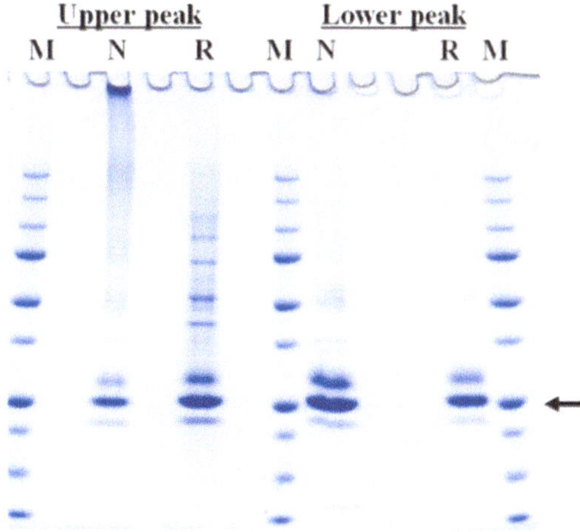

Fig. 2 Determination of the purity of the Foldon-mLIGHT protein by SDS-PAGE. *Lane M*: molecular weight standards, *N*: nonreducing condition, *R*: reducing condition. Upper peak contains Foldon-mLIGHT protein aggregates. Foldon-mLIGHT proteins were separated into three bands because it contains two N-glycosylation sites. Main band of the protein (*arrow*) had a molecular weight of 25 kDa under both the nonreducing and reducing conditions, which may be N-glycosylated at one site

4. Remove endotoxin from the purified proteins using an ActiClean Etox column. Freeze the purified Foldon-mLIGHT in small aliquots (100 μL) and store at −80 °C.

3.2.4 Determination of Foldon-mLIGHT Purity

The purity and structure of Foldon-mLIGHT protein can be estimated by SDS-PAGE and SEC, respectively.

1. Prepare a 10–20 % SDS-polyacrylamide gel and separate Foldon-mLIGHT protein by gel electrophoresis under reducing conditions in the presence of 0.1 M dithiothreitol or under nonreducing conditions. Stain the gels with SimplyBlue™ SafeStain (*see* Fig. 2 and **Note 12**).
2. Apply Foldon-mLIGHT protein to SEC to a Superdex 200 10/300 GL column in PBS or 10 mM sodium phosphate (pH 7.4) plus 1 M NaCl at a flow rate of 0.75 mL/min.

3.2.5 Foldon-mLIGHT Characterization by Cytotoxicity Assay

1. For maintenance of FM3A cells, use RPMI-1640 medium supplemented with 10 % FBS at 37 °C in a 5 % CO_2 atmosphere.
2. Seed the FM3A cells at 30,000 cells/well in 96-well plates in RPMI-1640 medium containing 1 % FBS and mouse IFN-γ (100 ng/mL). After incubation for 30 min at 37 °C in 5 % CO_2, add Foldon-mLIGHT with or without mouse LTβR-Fc (soluble LIGHT receptor) or a mouse IgG antibody as a control to the cell culture.

3. Incubate the cells for 2 days at 37 °C in 5 % CO_2. Add 10 μL of the Cell Counting Kit-8 reagent to each well, and incubate the plate for 1 h at 37 °C in 5 % CO_2. Measure the absorbance of each sample at 450 nm to measure viable cell number with 650 nm as a reference, using SPECTRA Max 340PC.

4 Notes

1. To disrupt *E. coli* cells completely, the lysate should be sonicated twice.
2. This step is important to reduce the impurities from inclusion bodies.
3. To maximize the yield, L-cysteine should be dissolved just prior to the denaturation process.
4. Although debris may be precipitated during the step, most of it is impurities and unfolded s-hLIGHT.
5. The solution should be immediately applied to a POROS® HS column. If the application of the solution is delayed, s-hLIGHT may be coprecipitated with impurities.
6. s-hLIGHT was eluted around 23 min as a single peak in our system; however, the elution time can differ for each system. Please optimize for your system.
7. We divided s-hLIGHT solution into two equal parts and applied each part separately to the column to optimize peak separation.
8. In our system, s-hLIGHT was eluted at around 45 min with a small front shoulder peak, and the front shoulder peak was eliminated. The elution time can differ for each system. Please optimize for your system.
9. s-hLIGHT was eluted at around 31 min as a single peak in our system; the elution time can differ for each system. Please optimize for your system.
10. s-hLIGHT may be precipitated during the buffer exchange step if D-PBS is used as a storage buffer. Therefore, we strongly recommend the use of 50 mM sodium acetate containing 0.15 M NaCl (pH 6.0) as a storage buffer.
11. Foldon-mLIGHT proteins were further separated and visualized as two peaks by size-exclusion chromatography using Superdex 200 16/60 column in PBS (pH 7.4). The upper peak was found to contain Foldon-mLIGHT aggregates.
12. Foldon-mLIGHT proteins were separated into three bands because they contain two N-glycosylation sites. The main band showed a molecular weight of 25 kDa under both nonreducing and reducing conditions.

Acknowledgments

The authors thank Drs. Ito T, Matsui H, Kurokawa T, and Taniyama Y for their kind guidance, constructive discussions, and encouragement throughout this work.

References

1. Mauri DN, Ebner R, Montgomer RI, Kochel KD, Cheung TC, Yu GL et al (1998) LIGHT, a new member of the TNF superfamily, and lymphotoxin alpha are ligands for herpesvirus entry mediator. Immunity 8:21–30
2. Tamada K, Shimozaki K, Chapoval AI, Zhu G, Sica G, Flies D et al (2000) Modulation of T-cell-mediated immunity in tumor and graft-versus-host-disease models through the LIGHT co-stimulatory pathway. Nat Med 6:283–289
3. Morel Y, Truneh A, Sweet RW, Olive D, Costello RT (2001) The TNF superfamily members LIGHT and CD154 (CD40 ligand) costimulate induction of dendritic cell maturation and elicit specific CTL activity. J Immunol 167:2479–2486
4. Shaikh RB, Santee S, Granger SW, Butrovich K, Cheung T, Kronenberg M et al (2001) Constitutive expression of LIGHT on T cells leads to lymphocyte activation, inflammation, and tissue destruction. J Immunol 167: 6330–6337
5. Browning JL, Sizing ID, Lawton P, Bourdon PR, Rennert PD, Majeau GR et al (1997) Characterization of lymphotoxin-alpha beta complexes on the surface of mouse lymphocytes. J Immunol 159:3288–3298
6. Harrop JA, McDonnell PC, Brigham-Burke M, Lyn SD, Minton J, Tan KB et al (1998) Herpesvirus entry mediator ligand (HVEM-L), a novel ligand for HVEM/TR2, stimulates proliferation of T cells and inhibits HT29 cell growth. J Biol Chem 273:27548–27556
7. Granger SW, Butrovich KD, Houshmand P, Edwards WR, Ware CF (2001) Genomic characterization of *LIGHT* reveals linkage to an immune response locus on chromosome 19p13.3 and distinct isoforms generated by alternate splicing or proteolysis. J Immunol 167:5122–5128
8. Rooney IA, Butrovich KD, Glass AA, Borboroglu S, Benedict CA, Whitbeck JC et al (2000) The lymphotoxin-β receptor is necessary and sufficient for LIGHT-mediated apoptosis of tumor cells. J Biol Chem 275:14307–14315
9. Chen MC, Hwang MJ, Chou YC, Chen WH, Cheng G, Nakano H et al (2003) The role of apoptosis signal-regulating kinase 1 in lymphotoxin-beta receptor-mediated cell death. J Biol Chem 278:16073–16081
10. Hikichi Y, Matsui H, Tsuji I, Nishi K, Yamada Y, Shintani Y et al (2001) LIGHT, a member of the TNF superfamily, induces morphological changes and delays proliferation in the human rhabdomyosarcoma cell line RD. Biochem Biophys Res Commun 289:670–677
11. Matsui H, Hikichi Y, Tsuji I, Yamada T, Shintani Y (2002) LIGHT, a member of the tumor necrosis factor ligand superfamily, prevents tumor necrosis factor-α-mediated human primary hepatocyte apoptosis, but not Fas-mediated apoptosis. J Biol Chem 277: 50054–50061
12. Hehlgans T, Mannel DN (2001) Recombinant, soluble LIGHT (HVEM ligand) induces increased IL-8 secretion and growth arrest in A375 melanoma cells. J Interferon Cytokine Res 21:333–338
13. Tsuji I, Matsui H, Ito T, Kurokawa T, Shintani Y (2011) L-Cysteine-enhanced renaturation of bioactive soluble TNF ligand family member LIGHT from inclusion bodies in *Escherichia coli*. Protein Expr Purif 80:239–245
14. Ito T, Iwamoto K, Tsuji I, Tsubouchi H, Omae H, Sato T et al (2011) Trimerization of murine TNF ligand family member LIGHT increases the cytotoxic activity against the FM3A mammary carcinoma cell line. Appl Microbiol Biotechnol 90:1691–1699
15. Meier S, Guthe S, Kiefhaber T, Grzesiek S (2004) Foldon, the natural trimerization domain of T4 fibritin, dissociates into a monomeric A-state form containing a stable beta-hairpin: atomic details of trimer dissociation and local beta-hairpin stability from residual dipolar couplings. J Mol Biol 344:1051–1069
16. Del Rio M-L, Jones ND, Buhler L, Norris P, Shintani Y, Ware CF et al (2012) Selective blockade of HVEM-BTLA pathway ameliorates acute graft vs host disease. J Immunol 188:4885–4896
17. Laemmli UK (1970) Cleavage of structural proteins during the assembly of the head of bacteriophage T4. Nature 227:680–685

Chapter 18

One of the TNF Superfamily Members: Bifunctional Protein, TNFR2-Fc-IL-1ra

Bo Xie

Abstract

Many anti-inflammatory agents have been exploited for the treatment of inflammatory diseases by targeting the most potent proinflammatory cytokines including tumor necrosis factor (TNF) and interleukin-1 (IL-1). Theoretically, simultaneous neutralization or blocking two important inflammatory mediators might achieve a synergistic therapeutic effect. A recombinant fusion protein, TNFR2-Fc-IL-1ra (TFI), was developed as bifunctional inflammatory inhibitor. TFI was able to strongly neutralize TNF activity and to antagonize IL-1 receptor in the cell binding inhibition assays, suggesting that TFI could be used as a bifunctional ligand with enhanced anti-inflammatory effect.

Key words Proinflammatory cytokines, Bifunctional protein, Synergistic therapeutic effect, Bioreactor, Affinity, Expression and purification

1 Introduction

Inflammatory reaction produced by the immune system is part of the complex biological response of vascular tissues to harmful stimuli, such as pathogens, damaged cells, or irritants [1]. A large number of cytokines have been recognized for their critical roles in inflammatory reactions after the onset of diseases [2, 3]. Among those recognized cytokines, TNF and IL-1 have been shown to play major roles in inappropriate inflammatory responses, and the neutralization of their activities has been effectively demonstrated in a large number of studies [4, 5].

As for the different modes of action between TNFR2 and IL-1 receptor antagonist (IL-1ra) and their different localizations in vivo, concurrent administration of TNFR2 and IL-1ra led to higher incidence of infection and neutropenia compared to the separate administration of these proteins. Theoretically, because of coordinated effects, simultaneous neutralization or blocking TNF-α and IL-1β may produce a synergistic anti-inflammatory effect in the treatment of inflammatory disorders [6].

Jagadeesh Bayry (ed.), *The TNF Superfamily: Methods and Protocols*, Methods in Molecular Biology, vol. 1155, DOI 10.1007/978-1-4939-0669-7_18, © Springer Science+Business Media New York 2014

The production of the fusion protein TNFR2-Fc-IL-1ra (TFI) that is recognized as an alternative drug for inflammatory diseases was carried out by the method of genetic engineering [7]. This study was designed to validate the feasibility of constructing a chimeric fusion protein with TNFR2 and IL-1ra dual domain, and to investigate its bifunctional TNF-neutralizing activity and IL-1 antagonizing activity.

2 Materials

1. DNA restriction enzymes: *Eco*RI and *Not*I.
2. Vectors: pMD18-T vector (Takara Bio Inc.), GC-rich vector pMH, pCDNA 3.1-TFI (Amprotein-China Inc.).
3. Primers: forward primer, 5′-cccccaagctggaattccaccatggagagagacacactcctgc-3′ (*Eco*RI site indicated by underlined bases), the reverse primer 5′-ggggaaaaaggtcgatgcggccgcttactactcgtcctcctggaagtagaatttggtg-3′ (*Not*I site indicated by underlined bases).
4. *E. coli* DH5α.
5. Cell culture and transfection reagents (Invitrogen Corporation) unless specified.
6. Complete RPMI 1640 medium: RPMI 1640, 10 % fetal calf serum (FCS).
7. Cell lines: L929 cell line (a murine fibroblast cell line sensitive to TNF-α cytotoxicity), murine D10 cells.
8. Cell count kit-8 (CCK-8) (Dojindo Mol Tech Inc.).
9. Genepulser Xcell electroporation system (Bio-Rad).
10. SP Fast Flow, Mabselect, and Sephadex G-200 columns (GE Healthcare).
11. SDS-PAGE apparatus, Western blot apparatus, Coomassie brilliant blue R-250, polyvinylidene difluoride (PVDF) membrane.
12. TBST buffer: 20 mM Tris–HCl, pH 8.0, 150 mM NaCl, 0.15 % Tween-20.
13. Blocking Buffer: TBST buffer, 5 % skim milk.
14. HRP-conjugated goat anti-human IgG, mouse anti-human TNFR2 antibody (Thermo Fisher Scientific).
15. CHO-S cells: Routinely cultured in DMEM/F12 (Gibco) or B001 medium (AmProtein).
16. AKTA purifier system (GE Healthcare).
17. Equilibration buffer A: 20 mM Tris–HCl, pH 7.4, 100 mM NaCl.

18. Elution buffer A: 100 mM Arg, 50 mM Gly-HCl, pH 3.5, 0.01 % Tween-80.
19. Equilibration buffer B: 100 mM NaAc-Tris, pH 4.5, 60 mM NaCl.
20. Elution buffer B: 100 mM NaAc-Tris, pH 4.5, 300 mM NaCl.
21. 100 mM phosphate buffer: Na_2HPO_4 (15.46 g/L), NaH_2PO_4 (5.83 g/L).
22. Microplate reader: Bio-Rad Model 3550.
23. Recombinant molecules: IL-1ra (Kineret), Enbrel (TNFR2-Fc).

3 Methods

3.1 Vector Construction and Transfection

1. Amplify the fusion gene coding for TNFR2 domain (NCBI accession M55994.1), Fc fragment (NCBI accession NM_001002274.2), and IL-1ra domain (NCBI accession X53296.1) from pCDNA 3.1-TFI by Taq DNA polymerase using the primers.
2. Clone the amplified DNA fragment into pMD18-T to yield pMD18-T-TFI.
3. Transform the pMD18-T-TFI into *E. coli* DH5α.
4. Excise the insert from the vector by digestion with *Eco*RI/*Not*I and ligate with *Eco*RI/*Not*I cut pMH [8] to generate the expression plasmid pMH-TFI.
5. Confirm the sequence of the insert in pMH-TFI by DNA sequencing.
6. Introduce pMH-TFI into CHO-S by electroporation using Genepulser Xcell.
7. Select the cells with stable expression of the fusion protein for protein expression (*see* **Note 1**) (Fig. 1).

3.2 Clones Screening and Recombinant Protein Expression

1. Transform the CHO-S cells with pMH-TFI by electroporation.
2. After transformation, seed the cells in 96-well plates at various cell densities ranging from 5×10^2 to 5×10^3 cells/well at 37 °C, 5 % CO_2.

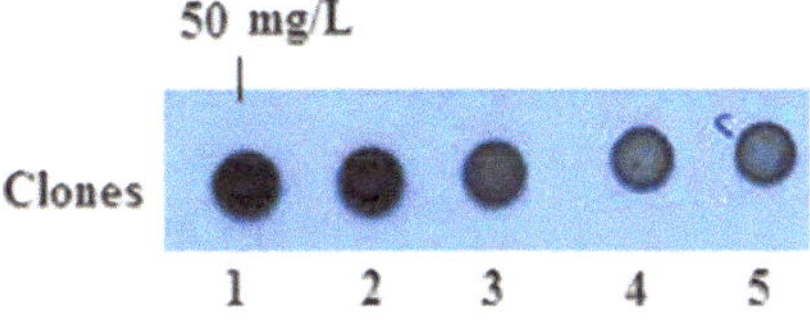

Fig. 1 Expression of TFI by CHO-S cells. Dot blot result showing the clones with high TFI expression levels

3. After 24 h of incubation, add fresh D/F12 medium plus 10 % of fetal bovine serum (FBS) containing 50 mg/L G418 to the cells and incubate the plate for 2 weeks at 37 °C, 5 % CO_2.
4. Enrich the single colonies through subculturing from 96-well plates into 24-well plates in D/F12 medium containing 100 mg/L G418 and subsequently into T25 flasks.
5. Collect the supernatants either from 96- or 24-well plates for initial assessment of TFI expression by ELISA.
6. Further culture the transformants with high expression of the target protein in 40 mL shake flasks and subsequently transfer to 5-, 50-, and 100-L bioreactors in sequence, with B001 medium for serum-free cultivation via fed-batch fermentation [9] (*see* **Note 2**).
7. After 13 days of cultivation, centrifuge the cultures at 5,000 × *g* for 15 min, collect the supernatants, filter through 0.22 μm membrane, and store at −80 °C for protein purification.

3.3 Purification

1. Perform all purification steps either on ice or in a refrigerator using an AKTA purifier system (GE Healthcare).
2. The purification protocol consists of consecutive chromatography steps on Mabselect affinity and SP fast flow columns (*see* **Note 3**).
3. Load the cultures onto a Mabselect column (50 mL; GE Healthcare) pre-equilibrated with equilibration buffer A, and elute with elution buffer A.
4. Pool the active fractions and load onto a SP fast flow columns (20 mL) pre-equilibrated with equilibration buffer B, and elute with elution buffer B.
5. Again pool the active fractions and concentrate before loading onto a Superdex-200 column equilibrated with 100 mM phosphate buffer at a flow rate of 0.5 mL/min.
6. Determine the purity of the purified protein with a Waters Breeze HPLC system (*see* **Note 4**).
7. Subject the purified proteins to SDS-PAGE and identify by western blot.

3.4 SDS-PAGE and Western Blotting

1. Resolve the protein samples in 8 % (w/v) polyacrylamide gel.
2. After electrophoresis, stain the gel with Coomassie Brilliant blue R-250 to visualize the protein bands (*see* **Note 5**) (Fig. 2).
3. For western blot analysis, transfer the proteins resolved in the gel to PVDF membrane.
4. Incubate the membrane with blocking buffer at room temperature for 60 min, and then incubate with mouse anti-human TNFR2 antibody (at 1:1,000 dilution in blocking buffer) at 30 °C for overnight.

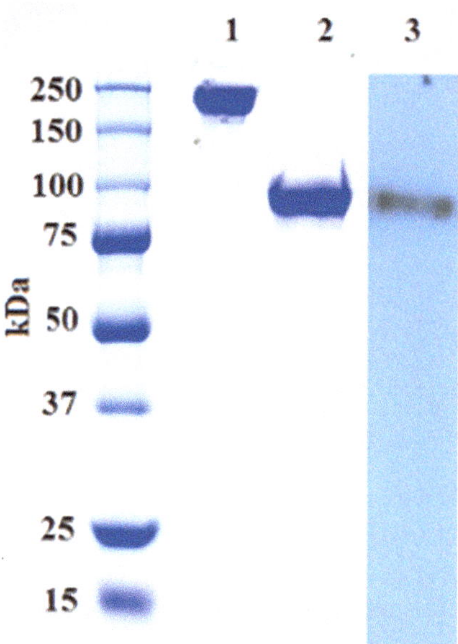

Fig. 2 Analysis of the TFI purity by SDS-PAGE and confirmation by Western blot. *Lanes*: *1*. nonreducing form; *2*. reducing form; *3*. TFI blotting with anti-human TNFR2 antibody

5. Wash the membrane three times with TBST buffer and then incubate with HRP-conjugated goat anti-human IgG antibody (at 1:5,000 dilution in blocking buffer) at 30 °C for 60 min.
6. After that, wash the membrane three times with TBST followed by detection using HRP (*see* **Note 6**) (Fig. 2).

3.5 TNF-α Neutralization Activity of the Recombinant TFI

1. Examine the anti-TNF-α activity of the recombinant TFI using L929 cell line [10].
2. Culture the L929 in complete RPMI 1640 medium in a 96-well plate at a density of 1×10^5 cells/mL at 37 °C, 5 % CO_2 for 3–4 h.
3. Then, add 10 ng/mL of TNF-α to the medium and add TFI to the cells at different concentrations (the initial concentration of TFI was 10,000 ng/mL and the dilution ratio was 1:3) and further incubate the plate at 37 °C, 5 % CO_2 for 24 h.
4. After that, add CCK-8 and leave for 2 h to determine the concentration of the effective dose (ED_{50}), and then read the plate at 450 nm in a microplate reader (*see* **Note 7**) (Fig. 3).
5. For control, use Enbrel (TNFR2-Fc) instead of the recombinant TFI.

3.6 IL-1r Antagonium Activity of the Recombinant TFI

1. Measure the IL-1r activity using a murine D10 cell proliferation assay with human IL-1β as the standard control [11].
2. Culture the D10 cells in complete RPMI 1640 medium in a 96-well plate at a density of 1×10^5 cells/mL at 37 °C, 5 % CO_2 for 12 h.

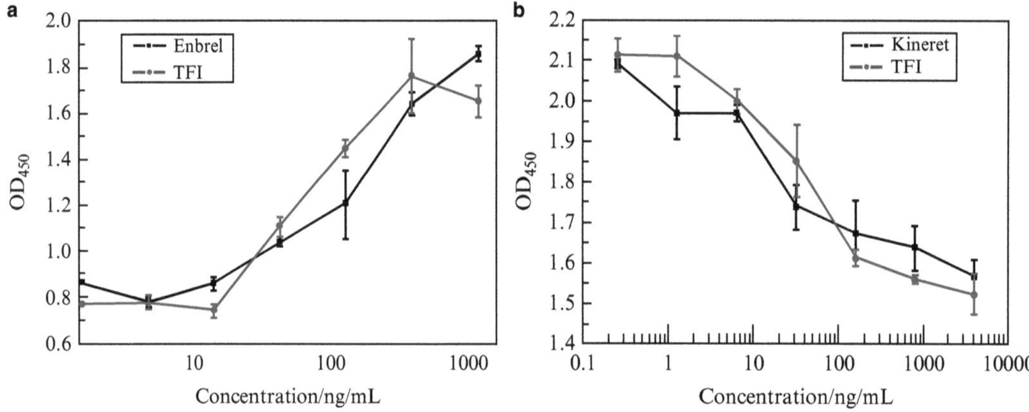

Fig. 3 Cytotoxicity activity assay of purified TFI. (**a**) TNF-α neutralization activity of TFI. The commercial drug Enbrel (TNFR2-Fc) was used as a control. (**b**) IL-1r antagonistic activity of purified TFI. The recombinant IL-1ra (Kineret) was used as a control. The half effective dose (ED_{50}) was determined according to the optical density at 450 nm (OD_{450}). The results were shown in mean ± SD of triplicate tests ($P < 0.05$). (Reproduced from [7] with permission from Springer)

3. Add different concentrations of TFI (the initial concentration of TFI was 10,000 ng/mL and the dilution ratio was 1:5) to the cells and then incubate the plate at 37 °C, 5 % CO_2 for 48 h.
4. After that, add CCK-8 to the cells and incubate the plate for another 4–6 h and read the absorbance of the plate at 450 nm in a microplate reader to determine the ED_{50} value (*see* **Note 8**) (Fig. 3).
5. For control, add recombinant IL-1ra (Kineret) instead of recombinant TFI.

4 Notes

1. A fusion gene, TNFR2-Fc-IL-1ra was inserted into the GC-rich vector pMH and introduced into CHO-S cells. After two cycles of transfection and selection, seven stable and high-level expression clones (50 mg/L in 24 h by dot blot analysis) were chosen from thousands of clones. As shown in Fig. 1, Clone 1 was applied for further study in serum-free medium.
2. Typically, Clone 1 was first cultured in T75 and propagated in serum-free medium B001 in 40-mL flasks for 2 weeks. Subsequently, the volume was expanded to 1 L. Finally, 5- and 50-L disposable bioreactors were used for product collection in fed-batch fermentation. Cell concentrations increase sharply and reach the highest level of 1×10^7 cells/mL at the fifth day, and then decreased with time.

Accordingly, the accumulation of TFI increases gradually at the initial stage, and although the cell concentration began to drop after the fifth day, its expression continues to increase. When the viability of cell decreases below 80 %, the expression level of TFI (determined by ELISA) reaches 180 mg/L (purification yield).

3. TFI products by CHO-S cells were purified from the culture supernatant using two steps of chromatography. TFI specifically adsorbed to the Mabselect column that was specific for the binding of the Fc fragment in TFI and the purity reached 85 % as determined by HPLC. Further purification of TFI by cation-exchange chromatography using SP Fast Flow column resulted in the purity of the protein greater than 98 %.
4. The endotoxin activity of TFI was less than 0.1 EU/mg.
5. The purity of TFI was assessed by SDS-PAGE, which showed a protein band of approximately 168 kDa under nonreducing conditions in line 1, and 84 kDa under reducing condition in line 2 (Fig. 2).
6. To confirm the antigenic activity of TFI, the protein resolved by SDS-PAGE was subjected to western blot analysis using anti-human TNFR2 monoclonal antibody. As shown in line 3 (Fig. 2), a single band was detected by the antibody, thus demonstrating its antigenic activity.
7. To test whether TFI could block the cytotoxic effect of TNF-α, TFI and TNF-α were applied together to L929 cells. The result showed that TFI could bind with TNF-α and neutralize its activity. Statistically, the effect became significant relative to baseline when the concentration of TFI was 40 ng/mL (Fig. 3a). More importantly, the ED_{50} of TFI was relatively smaller ((100 ± 0.93) ng/mL) than that of TNFR2-Fc (Enbrel, (180 ± 0.65) ng/mL) in linear range of (15–300) ng/mL ($P<0.05$).
8. In order to examine the binding activity of recombinant TFI to IL-1r, D10 cell proliferation assay was performed. As shown in Fig. 3b, the ED_{50} of TFI was (40 ± 0.71) ng/mL which was twice compared with that of recombinant IL-1ra (Kineret, (60 ± 0.94) ng/mL) in linear range of (1–4,000) ng/mL ($P<0.05$). The results suggested that TFI also has good antagonistic activity toward IL-1r in vitro [7].

Acknowledgements

This work was supported by 863 grant 2012AA02A406 and State Key Laboratory of Biochemical Engineering, Institute of Process Engineering, Chinese Academy of Sciences.

References

1. Ferrero-Miliani L, Nielsen OH, Andersen PS, Girardin SE (2007) Chronic inflammation: importance of NOD2 and NALP3 in interleukin-1beta generation. Clin Exp Immunol 147:227–235
2. Strober W, Fuss IJ (2011) Proinflammatory cytokines in the pathogenesis of inflammatory bowel diseases. Gastroenterology 140:1756–1767.e1
3. Miller AM, Mcinnes IB (2011) Cytokines as therapeutic targets to reduce cardiovascular risk in chronic inflammation. Curr Pharm Des 17:1–8
4. Vinay DS, Kwon BS (2011) The tumour necrosis factor/TNF receptor superfamily: therapeutic targets in autoimmune diseases. Clin Exp Immunol 164:145–157
5. Perrier S, Darakhshan F, Hajduch E (2006) IL-1 receptor antagonist in metabolic diseases: Dr Jekyll or Mr Hyde? FEBS Lett 580: 6289–6294
6. Bresnihan B (2001) The prospect of treating rheumatoid arthritis with recombinant human interleukin-1 receptor antagonist. BioDrugs 15:87–97
7. Xie B, Liu S, Wu S, Chang A, Jin W, Guo Z et al (2013) A novel bifunctional protein TNFR2-Fc-IL-1ra (TFI): expression, purification and its neutralization activity of inflammatory factors. Mol Biotechnol 54:141–147
8. Jia Q, Wu HT, Zhou X, Gao J, Zhao W, Aziz J et al (2010) A "GC-rich" method for mammalian gene expression: a dominant role of non-coding DNA GC content in regulation of mammalian gene expression. Sci China Life Sci 53:94–100
9. Eibl R, Kaiser S, Lombriser R, Eibl D (2010) Disposable bioreactors: the current state-of-the-art and recommended applications in biotechnology. Appl Microbiol Biotechnol 86:41–49
10. Baarsch MJ, Wannemuehler MJ, Molitor TM, Murtaugh MP (1991) Detection of tumor necrosis factor α from porcine alveolar macrophages using an L929 fibroblast bioassay. J Immunol Methods 140:15–22
11. Steinkasserer A, Solari R, Mott HR, Aplin RT, Robinson CC, Willis AC et al (1992) Human interleukin-1 receptor antagonist, high yield expression in E. coli and examination of cysteine residues. FEBS Lett 310:63–65

Index

Jagadeesh Bayry (ed.), *The TNF Superfamily: Methods and Protocols*, Methods in Molecular Biology, vol. 1155, DOI 10.1007/978-1-4939-0669-7, © Springer Science+Business Media New York 2014

N

O

P

Q

R

S

T

V

W

MIX
Papier aus verantwortungsvollen Quellen
Paper from responsible sources
FSC® C105338

If you have any concerns about our products,
you can contact us on
ProductSafety@springernature.com

In case Publisher is established outside the EU,
the EU authorized representative is:
Springer Nature Customer Service Center GmbH
Europaplatz 3, 69115 Heidelberg, Germany

Printed by Libri Plureos GmbH
in Hamburg, Germany